BIOTECHNOLOGY: PERSPECTIVES, POLICIES AND ISSUES

BIOTECHNOLOGY: PERSPECTIVES, POLICIES AND ISSUES

An International Symposium

Held at the University of Florida
Gainesville, Florida
June 1-4, 1986

Edited by
Indra K. Vasil

Graduate Research Professor of Botany
University of Florida
Gainesville, Florida

Sponsored by the Institute of Food and Agricultural Sciences, University of Florida; the IC[2] Institute, The University of Texas at Austin; and the RGK Foundation.

University Presses of Florida
University of Florida Press
Gainesville

UNIVERSITY PRESSES OF FLORIDA is the central agency for scholarly publishing of the State of Florida's university system, producing books selected for publication by the faculty editorial committees of Florida's nine public universities: Florida A&M University (Tallahassee), Florida Atlantic University (Boca Raton), Florida International University (Miami), Florida State University (Tallahassee), University of Central Florida (Orlando), University of Florida (Gainesville), University of North Florida (Jacksonville), University of South Florida (Tampa), University of West Florida (Pensacola).

ORDERS for books published by all member presses of University Presses of Florida should be addressed to University Presses of Florida, 15 NW 15th Street, Gainesville, FL 32603.

Library of Congress Cataloging in Publication Data

Biotechnology.

1. Biotechnology—Congresses. I. Vasil, I. K. II. University of Florida. Institute of Food and Agricultural Sciences.
TP248.14.B57 1987 303.4'83 87-14268
ISBN 0-8130-0883-2

CONTENTS

PART IV
GLOBAL, UNIVERSITY AND INDUSTRY PERSPECTIVES

PART V
COMMERCIALIZATION ISSUES

PART VI
SPECIAL LECTURES

APPENDIX

Francis Aloysius Wood
November 17, 1932—August 22, 1985

This volume is dedicated to the memory of the late
Dr. F. Aloysius (Al) Wood,
Dean for Research, Institute of Food & Agricultural Sciences,
University of Florida, Gainesville, Florida,
and
Chairman, Committee on Biotechnology,
National Association of State Universities and Land-Grant Colleges,
for his valuable contributions,
support, and understanding of the
value of biotechnology research and
development in agriculture.

PREFACE

The power, potential and promise of biotechnology has, in less than two decades, stirred worldwide interest in the application of this technology and its products for the improvement of plant, animal and human life. Universities, industries, governments, foundations, and international organizations are creating new opportunities for research, development, and applications of biotechnology. Early products of biotechnology, already available in the market place, are directed toward human and animal health. On the other hand, there are no major products of the agricultural (plant) biotechnology which are commercially available, although it is generally agreed that eventually the economic, social and political impacts of agricultural biotechnology products will far exceed those of human/animal biotechnology.

The euphoria of the early days of biotechnology, followed by serious difficulties in achieving the objectives, has gradually led to formulation of more realistic prospects and time tables. It has also become clear that substantial new basic research, particularly in the biology of plants, will be needed to exploit the full potential of these powerful techniques.

Even the current limited success of biotechnology has brought to attention important new concerns, problems and challenges. These generally are not discussed at the scores of technical meetings held each year in the United States and in other parts of the World. The perception of biotechnology in the public mind, and the concern for possible hazards to environment and life, have fueled considerable debate and discussion in communities, the Congress and the press. It was the late Dr. F. Aloysius (Al) Wood, who suggested that an international symposium be organized to focus attention on the many facets and problems of regulation, funding, priorities, intellectual rights, industry/university relationships, commercialization, societal and ethical concerns, international concerns and competition, and training. A distinguished international group of individuals representing academia, industry and governments was thus assembled to discuss and explore the present status of biotechnology, and to offer a forum for the exchange of views among academia, industry and government. I trust that the record of these discussions presented in this volume will serve to bring much needed attention to, and encourage discussion of, these varied and complex but critical aspects of biotechnology research and development.

The international symposium on "Biotechnology: Perspectives, Policies and Issues", was held June 1-4, 1986, at the University of Florida, Gainesville, Florida. It included a United States Senate Hearing on Biotechnology Commercialization by Senator Lawton Chiles (Democrat, Florida). This proceedings volume includes most of the material presented in the hearing as well as in the symposium (the complete program is included in the Appendix).

In the preparation of these proceedings for publication I have benefited greatly by the able and expert assistance of Lenie Breeze, who has cheerfully carried a heavy burden of work and provided the critical link between the authors, the press and myself. My gratitude also to Alicia Fry for editorial assistance in the preparation of the proceedings.

Indra K. Vasil

Acknowledgments

Financial support provided by the following organizations for the international symposium "Biotechnology: Perspectives, Policies and Issues" is gratefully acknowledged:

Institute of Food and Agricultural Sciences, University of Florida

The IC[2] Institute, The University of Texas at Austin

The RGK Foundation, Austin

Agricultural Division, The Upjohn Company and Asgrow Seed Company

Deloitte Haskins & Sells

E. I. du Pont Nemours & Company

Gainesville Area Chamber of Commerce

Monsanto Company

Purina Mills, Inc.

Talquin Corporation (Progress Center)

Westinghouse Electric Corporation

PART I
INTRODUCTION

Welcome Address

Marshall M. Criser, Jr.
President
University of Florida
Gainesville, Florida

Welcome to Gainesville, the University of Florida and the international symposium on "Biotechnology: Perspectives, Policies, and Issues." The University is pleased to host this symposium which is designed as a forum for leaders in business, government, and the academic world to look beyond research and scientific development to the implications and commercialization of biotechnology. The symposium is also designed to look at the related national policies on university and industrial relations, ethical questions, and economic opportunities.

The University of Florida is one of the largest and most comprehensive graduate research institutions in the nation. We are a land grant university in a state with special needs and with soils and subtropical climate more like other countries than other states.

The quality, quantity, and importance of the research conducted on this campus has already been recognized by the Carnegie Commission and by the Association of American Universities. The significance of the research has also been recognized by the amount of funding received for research. Here faculty in medicine, engineering, agriculture, pharmaceuticals, physics, business, biology, computer sciences, and other areas are already attracting more than $100 million a year in grants and contracts which have helped develop Bioglass, the Kalman filter, Gencor, Gatorade, and the first nuclear-pumped laser. From new varieties of peanuts and soybeans to toxin-synthesizing bacterial genes and synthetic insect hormones, from new technologies in CAD/CAM and robotics to developments in genetic engineering and breast cancer detection, University of Florida faculty are making major contributions to research.

What often begins as pure research is most likely to be conducted by the faculty and funded by the federal government. Applied research may be conducted by the faculty but is more likely to be funded by the private sector. For all the contributions from research faculty, and for all the support from the federal government, the real application and impact occurs when business and industry become involved.

One has to wonder what could be accomplished if the same emphasis had been placed on biotechnology as has been placed on electronic technology since the first digital computer was invented by University of Florida alumnus John Atanasoff. Although we have seen the cloning of plants from tissues, advances in cellular and molecular biology, embryo rescue and cell fusion, gene transfer and other genetic engineering, biotechnology has not received the same attention as electronic technology.

If we are to develop plants and animals that can withstand insects, diseases, and adverse weather, while producing greater nutrition and other products, we must dedicate the same kind of human and fiscal resources that we have given to information storage, retrieval, manipulations, and exchange.

The average American may be more concerned with losing weight than having a new grain in his morning cereal, but the possibility of developing public interest and support for biotechnology is probably better than ever before in our history.

Following the famine in Africa that was so dramatically depicted in our newspapers, on television, and through other media, public interest and concern has been widely expressed—from an outpouring of individual donations, to "Live Aid" and other mega events. In this country, the concern for the starving in Africa has been translated into helping our own food producers with "Farm Aid" and to helping our own hungry with "Hands Across America."

In spite of the criticism that Americans think that they can solve world hunger by sending a can of food to Ethiopia, most Americans and others realize that education and agricultural development are the real solutions to the long-range problems. It will not take much to convince Americans, and others throughout the world, that biotechnology holds the greatest promise for the future.

Perhaps we cannot develop a grain, or bean, or bird, or other food sources that can grow in the desert. But we may be able to develop plants to help stabilize the deserts or provide edible food sources from minute water supplies. Is it unthinkable that a plant might be developed that would pull hydrogen and oxygen from the air to provide its own source of water?

Creating what was once the unthinkable is the real challenge of biotechnology. This is not just for Africa. In this country, with our seeming abundance of food, we need to be concerned with producing food sources which yield more per acre, require less pesticides, and less fertilizers. We need crops that are more energy efficient and can be grown successfully without potential damage to the environment.

Twenty-five years ago, it was unthinkable to have a calculator that would operate on a battery smaller than a dime. It was unthinkable to have a portable computer with 640K memory, and now we have portable calculators and computers to help developments in biotechnology.

It took education, government, and business working together to achieve the developments in electronic technology, and it will take the same cooperation for advancements in biotechnology. John Atanasoff may have been the academician who invented the digital computer, but it took federal funding. And it still took IBM, Texas Instruments, Apple, and many others to make that technology available to individuals like you and me.

Nationally and internationally, we have a public interest that can be channeled to support biotechnology. One of our very real tasks will be to direct that interest, capitalize it, and use it for generations to come.

In Florida, the challenges of biotechnology have special meaning because our environment is different than in other parts of the country. The soybean that grows well in Virginia, Ohio, or Idaho is not successful in Florida. However, the varieties of soybeans and peanuts developed by our faculty in the Institute of Food and Agricultural Sciences are very successful here and in many parts of Africa and Latin America.

Florida's soil and climate have forced us to be concerned about pesticides, plant nutrition, and plant tolerance to heat, drought, salinity, and other factors. Our efforts to solve our own problems have led to international studies in agriculture that have benefited other countries, as well as our local farmers.

Because of our experiences and needs, because of the comprehensive nature of our University, and because of the special dedication of our faculty, the University of Florida has a special commitment to biotechnology.

I am pleased that you are here for this symposium and to explore research and scientific developments, their implications, and the related concerns about university and industry relations, ethics, and economics.

The State of Florida is also very committed to biotechnology. I have already mentioned some of the reasons, including our soil and climate. Another reason, perhaps rather obvious, is money. Agriculture has always been an important part of the state's economy, but economic growth in Florida has continued to outpace the national economy, primarily because of Florida's growth in high-technology manufacturing. Over the past decade, Florida has ranked seventh in the nation in terms of growth of manufacturing employment. Of Florida's top ten manufacturing firms, eight are in high-tech.

Since 1975, two-thirds of all new manufacturing jobs in Florida have been in high-tech manufacturing. Since 1975, high-tech employment has more than doubled. The number of high-tech and high-tech service firms has grown by 54 percent in the past five years. Today more than 27.3 percent of Florida's manufacturing jobs are in high-tech, compared with 18 percent nationwide.

While most of this high-tech growth has been related to machines, agriculture is still big business and the state recognizes the potential of biotechnology. Agriculture is still the state's second largest business, just behind tourism. One of every five jobs is in agriculture. Each year, the impact of agricultural production and processing on sales activities in Florida is nearly $16 billion. With a ripple effect, the total impact is nearly $43 billion, including the wholesaling and retailing of food but excluding restaurants.

Combine the state's strengths in agriculture and high-tech, and you begin to understand our enthusiasm for biotechnology.

Welcoming Remarks

Ray Iannucci
Executive Director
Florida High Technology and Industry Council
Office of the Governor
Tallahassee, Florida

Governor Bob Graham asked me to welcome you to Florida and to read to you his remarks. I am here representing the governor.

The governor is honored that Dr. Kozmetsky and the University of Florida's Institute of Food and Agricultural Sciences have chosen Florida as the location for an international symposium on biotechnology. You are in an exciting state and you are at an exciting university. Our state is committed to forging new frontiers in science and technology. There is interest in our industry, in our universities, and among our political leaders to support research to help Florida become a national and international leader in research.

Only a few decades ago what happened in Florida did not matter very much. To most of America, we were small, poor, and far away. Today what happens in Florida sets trends that effect the world. Florida is on the move and the University of Florida is on the move, as are our other universities. Florida has cast off the reputation as a "sun and fun" state and assumed its rightful position as a leader in innovation, higher education, and research.

As a major economic development goal, the governor has targeted biotechnology to be a major industry for Florida. This will require a solid academic foundation, a world class research faculty, top quality facilities, and instrumentation. Within our university system, the groundwork for this foundation already exists.

The governor's vision is to build upon this foundation and develop Florida into an internationally recognized source at the cutting edge of research in biotechnology. This will require commitment, tenacity, and foresight. The challenge before us in Florida is to develop our case to the legislature and how best to link the creative energies in the laboratories to the marketplace.

The state must examine its support of basic research. Research partnerships must be formed. As the state provides the financing for research and the scientists embark on a journey of discovery, the state policymakers must develop a tolerance for ambiguity and willingness to see our universities create an inventory of knowledge from which material progress will inevitably flow. An overwhelming majority of Florida's present industries are life science based—health care maintenance, citrus farming, food processing, fruit and vegetable farming, cattle and horse raising, forestry and commercial fishing. This makes advances in biotechnology all the more important to Florida's long-term economic well being, for while biotechnology represents a new industry in its own right, it is also a valuable resource of innovation that will help assure the future for many of Florida's existing industries.

An investment on the part of this state toward the development of biotechnology is an investment in keeping Florida competitive with the rest of the world, and in preserving the existing economic base of this state.

Welcoming Remarks

Charles B. Reed
Chancellor
State University System of Florida
Tallahassee, Florida

This symposium is a serious occasion, and biotechnology is a serious subject, but like all subjects in which there are elements of mystery, and even some fear, this subject is one which has often inspired humor. As in, "What do you get when you cross a saber-toothed tiger with a cobra? I don't know, but I wouldn't want one admitted as an undergraduate—even on a football scholarship." Nor would I recommend letting it out of the lab for a walk around the campus. And the reason I wouldn't goes to the heart of our discussions here today. The opportunities we have today, and the concerns of society about how we use those opportunities, are at the center of our discussions in this conference.

I am not a bioscientist, but my lack of advanced training in the field gives me one great advantage—the right to ask dumb questions. Those are always the hardest to answer. The questions I pose in beginning this conference are these: What do we mean by biotechnology? Is it just gene splicing? Would it include naturally occurring bacteria which can clean up an oil spill by digesting petroleum? Do we include the water purification functions of the sawgrass of the Florida Everglades?

The definition I like, and which I commend to you, is that of the Office of Science and Technology of the Congress. That definition is as follows: Biotechnology includes any technique that uses living organisms (or parts of organisms) to make or modify products, to improve plants or animals, or to develop microorganisms for specific uses.

The strength of this definition is its inclusiveness, for it permits us to come together from fields as diverse as molecular biology to that of practical agriculture, from laboratory work with DNA to the halls of our medical schools and hospitals, and from the realm of international scientific cooperation to the role of finance and banking in encouraging and sharing new technology.

In my brief words of welcome, let me offer one particular word of encouragement. This is a field which is uniquely positioned for the development of public/private partnerships, involving the institutions of the State University System of Florida. As chancellor, I want to go on record today as encouraging those partnerships, and promoting them. We need them, and we are prepared to make them work in Florida.

This is a major, international symposium. I am honored to be here. I am proud that our State University System, and in particular, our host institution, the University of Florida, have convened this symposium. And we are all very gratified to participate jointly with the sponsors: the Institute of Food and Agricultural Sciences; the IC² Institute of The University of Texas at Austin, Dr. George Kozmetsky, Director; and the RGK Foundation.

This symposium has essentially three goals: (1) to examine the state of the art of biotechnology; (2) to consider university and industry relations; and (3) to review economic opportunities, ethical considerations, and technology transfer issues related to the development and application of biotechnology throughout the world.

Clearly, many—but not all—of the challenges and opportunities which we face here today flow from our new-found ability to alter the genetic material in plant and animal cells. It is a simple statement, but it has not been a simple matter to acquire this new-found scientific competence.

For many thousands of years, the human race's collective body of knowledge remained virtually static or grew at such a slow rate, that adapting to change was not a great challenge. Today, coping with social impacts of technological change is a virtual industry, with popular authors, journalists, and therapists all finding their niche.

The fact is that we are simply creating new knowledge and a great deal faster than we have ever done before. It is a measure of the velocity of knowledge in our era that thousands of people around the world have lived long enough to encompass the arrival of piloted flight, global communications, and the creation of weapons capable of ending all life—in just one lifetime.

There are even a few people alive today who were born in the year 1884, the year in which Gregor Mendel, the founder of the science of genetics, died. If that seems like a long time ago, it was. In the same year, the French impressionist movement was in full flower, Brahms wrote his third symphony, and Mark Twain published "Huckleberry Finn."

Yet only 102 years have elapsed since Mendel's death, and in that time, human beings have learned how to elude barriers of time and space which had constrained us from our first appearance on earth.

With those lessons, as in the case of every significant technology—including the use of fire, the development of tools, and the ability to keep close track of time—deep social impacts have accompanied the development of new knowledge.

Just as the automobile, air conditioners, the airplane, the transistor, telephone and television systems, and the microprocessor chip have marked us and the society in which we live, so, too will biotechnology—in ways we will consider essential and desirable, and in ways we will probably dislike. We will be marked by this technology, because this is a technology which will be developed. We have no option of not going forward, for if we stop, others will go forward—perhaps with less care, with less caution, less restraint, and less responsibility.

There are two fundamental reasons for this imperative of development. First, it will be beneficial and profitable, and could provide one nation with a critical advantage over all others. And second, the curiosity of the human race does not allow it not to be developed. Our job is to see that this new technology is developed responsibly, and used appropriately.

As researchers, business people, and citizens, we have an obligation to do our work safely, and to produce a quality product. We must recognize that we are part of society, and that we are indebted to society for providing us the tools with which to work. We must recognize that we will be hold accountable for the fruits of our work. It is possible that we will find methods to feed the world and end hunger, protect our natural environment, cure disease, distribute prosperity widely throughout the world, and it is possible that we will develop things which are harmful, even dangerous.

Let me paraphrase a famous understatement of the significance of a new technology. It goes like this, "Some recent work by leading researchers, which has been communicated to me in manuscript, leads me to expect that biotechnology may become a new and important branch of knowledge in the immediate future. Certain aspects of the situation seem to call for watchfulness and, if necessary, quick action on the part of the administration."

All I have done is to substitute the word "biotechnology" for the word "uranium." The author of this assessment was Albert Einstein. The document was his famous handwritten letter to President Roosevelt. The date was 1939. Let us recognize the potency of biotechnology as we contemplate its benefits. Let us assess the risks at every stage, and act accordingly. May this major international forum serve as a signpost on the road to the responsible development of this new technology.

UNITED STATES SENATE HEARING

Excerpts from the United States Senate Hearing Conducted by the Honorable Lawton M. Chiles

Testimony from Dr. David Glass
Vice President

BioTechnica International, Inc.
Boston, Massachusetts

Thank you for this opportunity to present an overview of issues important to the commercialization of biotechnology. In presenting this overview, I would like to approach it from two different perspectives. The first is from the perspective of what the federal government can do to help industry with some of the issues and problems that I and the other speakers will be discussing today. As we will see, many of these issues are governmental in nature, so it is appropriate for there to be some governmental role in the responses to these issues.

The second perspective I would like to bring is that of a small company. You have just noted the importance of small biotechnology companies in this country in the commercialization of biotechnology products and, very often, we in the small companies have a slightly different perspective on things than some of the larger companies who are working in this field. My company, BioTechnica International, is such a small company. We were founded five years ago in Cambridge, Massachusetts, and we are a genetic engineering research company developing products primarily in two areas—agriculture and dental diagnostics, diagnostic kits for the detection of human periodontal disease and other dental diseases.

Our other interests in agriculture lie primarily in two areas—the development of improved microbial soil inoculants, such as Rhizobium, and the development of improved crop varieties themselves, primarily through the use of recombinant DNA techniques. Our first products here will largely involve herbicide-tolerant varieties of important crop plants. So I hope in coming from a small company, I can provide the small company perspective in my remarks this morning.

Although I, and other people, talk about biotechnology as an industry, it really isn't an industry so much as it is a collection of research tools and techniques, most of which having been developed in the past 10 or 15 years, and all of which promise to revolutionize the way living organisms are used for industrial purposes. Particularly, we are all aware of the great commercial benefit and the great benefit to society that biotechnology is expected to have in a number of industries, particularly human and animal health care, crop agriculture, and other areas such as food processing, chemical processing, and energy generation.

Most of these research tools were invented here in the United States within the past 10 or 15 years, stemming from basic research done in the 1970s at the nation's universities, largely funded by federal research grants, and indeed, it was this large influx of federal research funding which was largely responsible for the United States holding the clear uncontested lead in biotechnology as the 1970s ended, and that is both in the basic research area and in the emerging commercialization of biotechnology that began around the end of the 1970s decade. However, as we have moved from the seventies to the eighties and the technology has begun to move to the marketplace, the United States has seen its lead slowly eroding to the point today where we can say that we only hold a fragile lead over our international competitors. This is the first issue that I would like to raise today, the issue of international competitiveness and how the United States might maintain its lead. While this subject is often mentioned by industrial speakers in settings such as these, it is difficult to over-exaggerate the importance of this issue for the industry and for the country. We cannot let happen to biotechnology what has happened, as you noted to so many other areas of technology in the past 20 years, areas that were largely invented in the United States, but would end up being primarily commercialized by companies in other countries.

I would like to mention two particular manifestations of this, one a specific case and one a general case, where present governmental policies either are, or run the danger of, contributing to this erosion of the United States' lead in the technology.

A specific example is the situation regarding the export of unapproved pharmaceutical products. Current United States law prohibits a company to export a pharmaceutical product before it has been approved by the Food and Drug Administration. Because the FDA review procedures are so thorough and so lengthy, most drug products receive approvals in foreign countries before they receive FDA approval here in the United States. Companies not wishing to lose a year or two or more of marketing time will find a way to manufacture these products outside of the country, so that they can supply markets where they receive regulatory approval. When this is done, American jobs are lost and American technology is transferred overseas, neither of which are particularly happy occurrences.

I might add that this circumstance affects biotechnology companies in particular because most small biotechnology companies need larger corporate partners to market pharmaceutical products, and this state of affairs would prompt them to look for foreign collaborators rather than collaborators in the United States.

As I am sure you know, there have been several bills introduced into Congress to alleviate this situation, one of which was recently passed in the Senate about a month ago. This bill would allow certain pharmaceutical products to be exported to certain countries once they have received regulatory approval from those countries. It is very important that this bill be passed in the House and signed by the President. It is a very important bill for that sector of the biotechnology community developing products in the pharmaceutical area. Not only is that the largest sector of the biotechnology industry, but it also could well be the one promising the most benefits to society in terms of novel therapeutic products that treat cancer and other deadly diseases.

A more general problem regarding U.S. competitiveness is the overall regulatory structure and climate in the United States. The United States, of course, is generally known to have perhaps the toughest regulatory regime in the world. Now oftentimes this is absolutely necessary for the adequate protection of the public health and safety and for the protection of the environment, but it is important to note that when overly strict regulations are applied to an emerging technology at an early stage in that technology's growth, and in addition, at early stages of product development, you run the real risk of stifling the development of the industry. I think this will be most keenly felt in the areas of agricultural biotechnology, as I will describe in a moment.

In general, I would like to point to the regulatory climate and the regulatory regime as the second major area of importance to the commercialization of biotechnology, and of course, a speaker later this morning, Mr. Korwek, will go into this issue in more detail.

As I have said, this situation will primarily affect products in agricultural biotechnology, and this is primarily because of the way that agricultural products are developed in this country. That is not just biotechnology products, but agricultural products in general. Companies developing agricultural products such as a new pesticide or a new strain of crop plant need to do intensive testing on an iterative basis in the field to evaluate what sometimes are many potential product candidates to narrow it down to the one or two that are going to be commercially viable. Greenhouse testing is not a 100 percent accurate indication of what is going to happen in the field. This is the reason why companies have to do extensive field testing. As I have said, in an iterative type of way, you take what you learn the first year and then go back to the field the second year and so on, repeating the process until you have come down to the candidate that you know will work. What this implies, at least in biotechnology, is that many of these early field tests that have been lately proposed and have been the subject of some controversy, are actually highly preliminary product prototypes. Some of them aren't even product prototypes at all, as much as they are test cases or preliminary products, so by imposing stringent regulatory requirements for field testing on these types of tests runs the risk of hindering innovation and stifling of development of agricultural products from biotechnology. This, of course, is not to say that there shouldn't be any regulation of field testing, and indeed, one of the challenges that faces us is to come up with a regulatory regime for field testing that will allow products to be field tested and eventually to allow products to come to the market while still protecting the environment, and while still insuring the public that health and safety considerations will be observed.

Currently in Washington, we have what amounts to a logjam right now with the government not quite knowing how to regulate these initial field tests. The result being that field tests, at least of microorganisms, are not getting done. A colleague of mine, Dr. Ralph Hardy, has made a proposal for a potential solution to this logjam which I would like to briefly mention here today for possible discussion later on. He has proposed that the federal government establish publicly-owned, professionally-managed field testing sites perhaps utilizing existing agricultural testing stations run by the government. At these stations, companies would come using their own personnel to conduct the tests, but they would be overseen by personnel from the public sector who have the responsibility for running the field site, of course, but also for overseeing the test. In this way, the government and the public can be assured the tests are done according to the proper protocols and with the proper monitoring, while still allowing the test to go on in a timely basis.

Let me close off talking about regulations by simply noting, as I did earlier, that we are not saying that there should be no regulations. Everyone in industry believes that there should be regulations in biotechnology, particularly in the environmental area. All that industry is asking is a stable, predictable regulatory regime by which we will know in advance what we need to do to get our products tested and out to the market, what tests we need to do, how long those tests will take, and what it is going to cost. Knowing this, we can incorporate all this information into our strategic planning and this will be a desirable result for the industry. I think clearly the administration and the Congress are moving towards this, but sometimes the pace of this movement seems to be a little slow.

Let me move on to some other areas that are of importance to the commercialization of biotechnology. Our next speaker, Congressman MacKay, will be talking about issues of intellectual property protection. This is clearly of great importance to the industry because no company will commit the large amounts of time and money to develop a product without knowing that they will have adequate protection for their products under the nation's intellectual property laws, primarily the patent system. The biotechnology industry has been heartened in recent years by two decisions, one the famous 1980 Chakrabarty decision of the Supreme Court which held that microorganisms could be patented, and more recently the decision by the Patent Office Board of Appeals in the Hibberd case about six months ago that plant materials can be patented under the general patent statutes. They are encouraging and lead us to believe that we will be able to have adequate patent protection for our products and biotechnological inventions; however, let me insert the caveat that until there is more experience with biotechnology patents in the patent office and in the courts, more case law, perhaps even more litigation, it will not be clear to what extent biotechnology patents will be enforceable, what scope will be available for them, and frankly, whether or not patents will have any value in such a rapidly moving field of technology as biotechnology. However, right now we are all placing a lot of stock in the patent system. We are all spending a lot of money trying to patent our inventions, and for this reason, the U.S. Patent Office needs to be strong and needs to be able to deal with these applications as they come in. As I am sure you know, there is now a considerable backlog of patent applications at the Patent Office, currently running about a year-and-a-half between initial application and first office examination, with issuance a year or two after that. The Patent Office needs more examiners and they need better-trained examiners in the area of biological sciences. This should also be a responsibility of the government to make sure that the Patent Office is well-equipped to deal with these issues.

Let me say a few words about potential legislative approaches to the patent system, knowing that Congressman MacKay will probably spend some time dealing with these issues. There have been a couple of potential new areas of legislation that have been mentioned and proposed in Congress. I will say a few words about two in particular. The first is patent term restoration, whereby a company is compensated for lost marketing time under its patent while it is waiting for marketing approval from a federal regulatory agency. Patent term restoration for certain pharmaceutical products was passed two years ago by Congress and there are now several bills pending that would apply the same type of patent term restoration to agricultural products. Those of us in the agricultural sector would welcome agricultural patent term restoration.

Another major issue is process patent infringement. This is a situation where a company holds a U.S. patent to manufacture a product, but does not hold foreign patents on that process. A competitor infringes the patent overseas and then comes back to the United States to sell the product. Current law gives the U.S. company very little recourse against such infringement. There are several laws pending in Congress that would make this act an infringement, just as if the infringement occurred in the United States. This type of bill is extremely important for the biotechnology community, particularly because so many of our patents will be process patents covering manufacturing processes, rather than product patents, so this too is a piece of legislation that we would like to see passed.

Another area that is of great importance to the biotechnology industry, as well I am sure to a great many people in the audience today, is the issue of basic research and the continued funding for university programs in molecular biology, genetics, and other related areas. We have already noted that the biotechnology industry owes its start and owes a great debt to the large amounts of federal funding and the excellent basic research conducted in this country in the 1970s. We all continue to be dependent on future research at American universities as the industry develops. By this I mean not only basic research by which we learn more about life and life processes, but also the kinds of applied research sometimes funded by corporations, sometimes funded by the government, which actually lead to potential product opportunities for small companies and large companies alike. We are also dependent on a third area of university research which has become increasingly important in today's regulatory climate, and that is research in the areas of risk-assessment, particularly environmental or ecological research which can and will be used to support the kinds of risk-assessments the companies will need to provide support for the registration and approval of their products down the line.

All these types of research are important to be done at universities. Some of the larger companies in our business, the established companies, are of course, funding a great deal of research and will continue to do so. However, funding significant amounts of research is far beyond the abilities of companies such as my own, and I think it is safe to say in this country that funding of basic research remains the responsibility of the government. For the biotechnology industry, it remains a very important responsibility and function of the federal government.

Let me also just mention that in general, the relationships between industry and the universities in this country remain important. I believe some of our later speakers will discuss in more detail some of these issues concerning such relationships, including licensing agreements and commercial arrangements. We are all dependent on the universities for the training of our future employees, the scientists, engineers, and managers, that we will need as our companies grow and develop. Again, I think we will hear more on this topic from our speakers later this morning.

Another issue of great importance is the general issue of economic development. That is, how can the government help us to foster a favorable climate for the development of biotechnology companies? Let me just say two brief words about this. In general, one thing that would be very helpful would be governmental programs that would help foster a favorable climate, economic and otherwise, for the development of companies and for the expansion of companies. In this country, we have seen models in certain state governments setting up biotechnology programs or biotechnology centers designed to provide economic incentives, tax breaks, etc., for companies to relocate in these states, expand their facilities into these states, or set up manufacturing facilities. We have also seen foreign countries setting up the same types of programs. Particularly some of the European countries are establishing the same sorts of programs and are very aggressively contacting U.S. companies, primarily small biotechnology companies, to set up their manufacturing facilities overseas. This of course, gets back to the international issues I mentioned earlier. It is in the best interests of the U.S. government not to allow foreign countries to entice our companies away from us, so the federal government should consider what types of programs it might be able to do along the lines that some of our own states have been doing.

The other issue in the area of economic development and capital formation is what types of financial incentives are appropriate for an industry such as biotechnology. Here I will only note that the more traditional means of financial incentives, such as R&D tax credits, are not particularly useful for companies such as ours who are not paying taxes because we are not making a profit yet, so the more traditional areas of financial incentives might not be as applicable to our industry as it might be to other industries. Industry and government together should work towards finding creative, innovative ways to provide incentives for the types of longer term, higher risk ventures that are characteristic of the biotechnology industry.

Let me close by discussing one last issue that, in the end, will probably prove to be more important than all these other issues combined, and that is the issue of public acceptance of the biotechnology industry and of our products and of our activities. In many cases, the general public will be the ultimate consumer of biotechnology products. In other instances, and really in all instances, the public, through their elected representatives and through the appointed members of the federal government, have the responsibility for ensuring the safety of our products and assuring that everything we do is done according to proper procedures. Clearly, there is a lot to be done in the area of public information, leading toward public acceptance of what biotechnology does. Several areas are worth pointing out here. Certain sectors of the public still have religious or moral concerns about what we do and about the effects of biotechnology. They will need to be addressed in the years to come. In addition, there are a whole host of ethical issues concerning the way the research is carried out, the impact of the research and its products on the economy, and many other factors. All of these need to be considered.

Then finally, there are the health and safety concerns and the environmental safety concerns which today are very prevalent in the public, and which, of course, are of extreme importance as the industry moves from the laboratory to the marketplace, and particularly as the industry moves from the laboratory to the field in agricultural field testing.

The solution to all these problems must come through considerable efforts of public education and public information dissemination. This largely has to come from the industry itself, as we all have to be much more forthcoming with the public, with the trade press and the lay press, the popular press, the Congress, and with other members of the public sector. Industry is committed to doing this and, through our trade associations, have lately begun such public information activities. But we do ask that wherever appropriate, and whenever possible, the government and other members of the public sector help us in this regard, particularly from the university sector, where a few words from a qualified academic expert can sometimes be very useful in putting things into a proper perspective.

I hope that in this brief time I have been able to provide an overview of important issues for commercialization of biotechnology. I am sure that many of the points I have discussed will be discussed in much greater detail by the speakers and witnesses who follow. I will be glad, Senator, to answer any questions that you might have either now or later.

Testimony by Dr. Jefferey Burkhardt
University of Florida

I appreciate the opportunity this morning to address you on the issue of ethics, social values, and the new biotechnologies. The new biotechnologies, unlike many other techniques or technologies, present us with the opportunity to assess our values and scientific and institutional commitments before we have plunged head-long into adoption, diffusion, and commercialization.

Many mechanical technologies are already in place. We are in a position to assess their impacts, but only after they have occurred. We know, for example, that the mechanical harvester has had profound social and economical effects. Some of these have been for good, some for ill. The present state of biotechnology allows us to reflect on possible effects before they occur. Hopefully we might prevent many, if not all, ill effects from occurring.

The new biotechnologies also present us with the opportunity to reflect on what our values are—that is, what we as individuals and as a society find really important. What are our values regarding science, technology, our institutions, and the environment? What value conflicts do these technologies present us with or exacerbate? What problems might they present us with in the future?

My comments today are based on a study I conducted from 1983 to 1985 concerning the ethical and value issues surrounding the development and applications of the new biotechnologies in agriculture, particularly regarding plant improvement. The study was funded by the National Science Foundation's Ethics and Values and Science and Technology Program. The other investigators and I personally interviewed over 100 scientists and research administrators across the country in both the public and private sectors. We sought to ascertain exactly what was being done with biotechnology in plant improvement, as well as what impacts biotechnology is having and will have on the practice of science and on the broader society.

My role as ethicist was to frame questions concerning values, interpret and clarify the answers received from interviewees, and most importantly, draw from our interviews the logical implications regarding value commitments, institutional changes, and future prospects for biotechnology. I would like to briefly summarize concerns voiced by scientists I interviewed, as well as some of the implications of these concerns.

There are three distinct, though interconnected, sets of ethical issues which can be gleaned from discussions with those in biotechnology. These three ethical problem areas are professional ethics, the changing mission of public institutions, and responsibilities regarding the environment, future generations, and the future of science.

The professional ethics problem area includes issues related to the proper relationship between the practice of science, professional responsibility, and personal integrity. Some of the cases that were raised as being of major importance to bench scientists were the following. Biotechnology is "in." Since the passage of the Plant Variety Protection Act, and since the 1980 Chackrabarty decision, plant varieties and novel life forms can be patented. There is, understandably, a race among scientists to be the first to introduce a bioengineered product to the public, or in the case of the private sector, being the first to develop a patentable product for the market. This can conceivably cause sloppy science, even dangerous science. The concern is that long-standing cannons of professional responsibility, including commitment to careful, rigorous, checked and re-checked scientific results, might give way to fast and loose science. In the long-run, shoddy science or faulty products would undoubtedly be excised from the scientific community or from the market. The concern is, however, about the short-run.

New university/industry relations, such as privately funded biotech institutes on university campuses, also cause concern. Proprietary rights, which Congressman MacKay spoke about this morning, of corporate funding sources to biotechnologies can conceivably impede the free flow of scientific information, even among public sector researchers, thereby decreasing the rate of discovery, replication of results, and so forth. Again and again, scientists I spoke with voiced difficulties with sharing information. If scientists can't legally communicate with each other, progress in science is hampered.

The potential for conflicts of interest was also repeatedly noted. Scientists and administrators have expressed concern that as private funds are increasingly available, while individual public sector scientists retain a university component in salary or research support, situations where the scientist is serving two masters may arise. More importantly, however, is the potential for scientists to use information or research results for personal gain, whatever the funding source. Both scientists and administrators know the danger of violations of employment contracts and role responsibilities. Certainly the potential for conflicts of interest or contractual violations existed long before the emergence of biotechnology; however, the sheer amounts of monies to be made from biotechnological products and processes have increased the opportunities for such conflicts.

The National Academy of Sciences and the National Association of State

Universities and Land Grant Colleges have attempted to prevent such problems through legal mechanisms, making as specific as possible the nature of public/private agreements, detailing what all parties can expect. How individual scientists or administrators follow the letter or the spirit of the law remains to be seen. Even with some legal instruments available or in place, those involved in biotechnology continue to express concern about honesty and conflicts. Some go so far as to blame the potential for dishonesty or conflicts on the system. This leads me to the second problem area.

The administrators I talked with expressed concerns about ethics and values in their institutions. Among the situations cited were the following. Funding cut-backs for public research or at least greater competition for scarce resources have led public research administrators to seek private assistance and also to look to shorter term results for R&D projects. Private firms have looked to the university as a source of scientific talent with labs and research assistants already there. Many administrators raise the question of whether new institutional arrangements such as university/industry contracts or privately supported biotech institutes on university campuses may too drastically shift the historical division of labor between public and private research agendas, perhaps threatening the kinds of research traditionally done in the public sector.

Questions were raised about the public sector's ability to serve a broader constituency such as small farmers, farm labor, backyard gardeners and the like. While most scientists and administrators believe that these traditional groups will be served over the long-term through biotechnology, the short-term effects are what worry them. Interestingly, many private research administrators express similar worries.

One specific problem raised in regard to the public sector was that some programs or departments which, because of their discipline, cannot receive research support related to biotechnology might suffer. Among them, for instance, social sciences, such as rural sociology and agricultural economics, and perhaps even agricultural engineering. Since these programs or departments have historically served a broad-based constituency, including rural communities and small farms, de-emphasis of these programs in the overall research agenda may have negative social effects.

The issue here is the mission of public research. Most of the scientists I talked with, and administrators as well, recognize that the mission of a public institution will change as client groups change. In Florida, for example, the rural/urban shift means food and agricultural research must now pay closer attention to the needs of urbanites. Biotechnology can assist them in development of techniques and products serving urbanites. However, in the minds of many scientists, administrators, and client groups, this research should not lead the way toward the more rapid decline in rural clientele by undercutting research favoring small farms or rural communities.

The last problem area is more amorphous, but it was raised independently by scientists and administrators that I talked with. The last issue is what we will leave posterity. There is an overarching sense that with any new scientific discovery, potential exists for tremendous beneficial effects. There is also a sense that ill effects are also possible. Most people were aware of Rifkin's lawsuits raising the spectra of the potential long-term environmental harms which could result from the release of bioengineered organisms into the environment. Many felt that whatever the merits of such cases, environmental effects cannot be lightly regarded. But interestingly, part of the problem people noted was that neither the public, nor those involved in biotechnology, were fully aware of the risks or lack of them associated with biotechnology. To my mind, this raises a question about the extent to which those involved in risk-assessment or decision-making about biotech might be ethically obligated to inform the public and others about those risks or lack of them. If we have adequate risk-assessment, then, let us inform those concerned about biotechnology so as to allow us to continue sound research in this area. If we don't have adequate risk-assessment, more or better is probably necessary.

One final issue was raised. This is the issue of what sort of science we will leave future generations. The new biotechnologies are the result of the long-standing historical trend toward mechanistic reductionism, the review that all phenomena can be explained only in terms of the complex working of increasingly smaller and smaller parts. No one has disputed, nor can dispute, the profound successes that this philosophy of science has had across the board in terms of products, processes, and the amelioration of the human condition. Nevertheless, concern was raised. If we become so enamored with biotechnology as to systematically ignore more holistic sciences or methodologies, or if we commit resources in such a way as to destroy the base for broader research, or if we nearsightedly come to view all facets of human life only in terms of the most microbiological processes, we may be doing posterity a disservice. We may be blatantly violating ethical obligations to the future.

I have spoken on three problem areas where ethical and value issues have, and will, continue to emerge in regard to biotechnology. In my judgment, this implies that further research is necessary on the ethical and value dimensions of biotechnology, not only in agricultural resources, but also in medicine, pharmaceuticals, and the like. A more complete awareness of what we as individuals and we as a society think is important, and what conflicts among various value systems exist, can only aid in decision-making. Individual research by philosophers and social scientists, but more importantly, interdisciplinary and joint research is clearly implied. Every sponsor of biotechnology research and development projects should perhaps have a values or ethics component built in. Many of the risks or impacts associated with biotech are simply not known at this time, but without broad and forward-looking perspectives introduced into the very core of our research, we may come to know only too late the full consequences.

Even with the proliferation of research on ethical and value implications, additional actions may be necessary. Education for values awareness is another need. Decisions are made by individuals. Many ethical and value problems are undoubtedly resolvable in the consciences of individual decision-makers. The presumption is, however, that scientists and decision-makers know how to see and how to address an ethical or value problem when they fall upon one. Here at the University of Florida and at universities across the country, we have begun to address the mistake in this presumption. Courses are offered where future scientists and decision-makers can explore ethical problems they will face in their professional lives. In these courses, future leaders are provided with logical, conceptual, and analytical skills appropriate to perceiving, facing up to, and resolving where possible, ethical issues such as conflict of interest, scientific honesty, and the goals of our institutions, including the future of science.

Despite their importance, additional research and college courses in scientific and professional ethics are not enough. There are thousands of scientists and administrators who cannot wait for future research results or for present students to join the ranks. The value and ethical issues involving biotechnology are presently here. This is why presently existing means may be used to draw attention to and help address ethics—in colleges of agriculture, in particular.

Extension may be the vehicle for providing this necessary service. Extension programs focusing on professional ethics for private sector scientists, for example, may help fill the gap in ethics awareness. Programs on managerial ethics for public and private administrators might serve a similar purpose. In general, values and ethics awareness programs are necessary both to explore, as well as to explain, the benefits and risks of biotechnology. These would enable more scientists and administrators to better assess the complete aspects of biotechnology.

Others here today are, of course, better qualified than I am to speak about existing proposed legislation. I will, however, offer three modest suggestions related to the concerns that have been raised regarding professional ethics, institutional missions, and our responsibility to posterity. (1) Continued public funding for other kinds of research, including social science and more holistic methodologies, may help address impacts before they occur. (2) Continued risk-assessment utilizing the best results, not only from environmental and ecological sciences, but also from the social sciences, may give the broader perspective necessary for wise decisions regarding biotechnology. (3) In a different vein, biotechnological review committees at either national or university levels which include an ethics and values component may both protect against shoddy research, as well as provide greater legitimacy for well-conceived, soundly carried out biotechnological research and development.

In summation, then, scientists and administrators I talked with believed that the new biotechnologies are having, and will have, great social benefits. They believe that the new biotechnologies may also carry great social risks. I agree with both points. Concerns have been expressed about (1) the impacts on professional values; (2) the mission of public research; (3) the environment; (4) future science; and (5) effects on future generations. Further research on these impacts is necessary, as is continued and improved environmental risk-assessment. Education of both professionals and the public as to the prospects and promises of biotechnology is also necessary. Awareness about values and ethical conflicts brought about by the biotechnology revolution is also necessary. Future legislation should support the work of social sciences and humanists in regard to this area of concern. Where possible, future legislation should use, as a guide in decision-making, the results of such research. In doing so, we may be able to guarantee, as far as possible, that the broadest range of human and environmental values will be considered and respected.

Testimony from Congressman Buddy MacKay

I would like to first thank Dr. Kozmetsky and the other persons who are sponsoring this conference, and I would like to thank Dr. Glass. I wish my colleagues could have heard that presentation. I think it would be certain that the Congress, when it acts, would act in a knowledgeable way. My comments will not be as broad as Dr. Glass', although I have to say that even when you talk about a "narrow" part of this expanding issue, you are talking about incredibly broad matters. President Lincoln, in the beginning of a very famous speech, said, "If we could first determine where we are and whether we are attending, we could better decide what to do and how to do it." From my standpoint that threshhold question is one that has been inadequately addressed and I would like to spend some time talking about that.

The spring issue of *Foreign Affairs* magazine has a major article by the economist, Peter Drucker, in which he poses the thesis that we are in a changed world economy. I would like to read three quotes just from the first page. They set forth the thesis of my remarks today. He says,

> The talk today is of a changing world economy. I wish to argue that the world economy is not changing. It has already changed its foundations and its structure and in all probability the change is irreversible. It may be a long time before economic theorists accept that there have been fundamental changes and longer still before they adapt their theories to account for them. Above all, they will surely be most reluctant to accept that it is the world economy in control rather than the microeconomics of the nation's state on which most economic theory still exclusively focuses. Practitioners, whether in government or in business, cannot wait until there is a new theory. They have to act and their actions will be more likely to succeed the more they are based on the new realities of the change in the world economy.

That provides emphasis for the thesis of my remarks. The emerging issue for the rest of this century is the need to compete in the world-wide marketplace. Dr. Glass touched on that in his overview. The new economy is a world economy. It is an economy characterized by a quickening pace of change, ceaseless innovation, and unprecedented flows of capital. The energy source fueling the new economy is not oil or coal, but knowledge and ingenuity. Raw material is not steel, but human brainpower. I will come back to the energy source in a minute.

Against that background, it is clear that intellectual property rights as an issue has changed from a dry, esoteric issue to a central issue, one of the key issues determining whether America and American science will be competitive in the world marketplace.

Coming back to the energy source, once again quoting President Lincoln, he made this statement that I think is remarkable, "The patent system adds the fuel of self-interest to the fire of genius."

We are accustomed to that idea and America has a good system for protecting intellectual property rights. It is somewhat in turmoil today as Dr. Glass pointed out, and I think he pointed out in a thumbnail manner the risks that are faced in our system by innovators, inventors, and small firms particularly. A small firm has got to set aside a portion of its budget for litigation and this system as it seeks to catch up with this wholly new idea that life forms and living organisms can be patented and that people can have property rights in that, part of it is going to be worked out in the legislative branch in the Congress, part of it is going to be worked out in the courts. Be that as it may, and given the turmoil and accepting as an excellent thumbnail sketch what Dr. Glass has said about the places where the questions remain to be resolved, internationally there is not even agreement that there should be property rights and ideas. There is not even agreement internationally that there are intellectual property rights much less property rights and life forms and living organisms.

Some countries believe, and have the tradition, that ideas are the common property of mankind. The internationally recognized standard is that foreign intellectual property rights should be treated the same as domestic intellectual property rights, so some Third World nations have systematically weakened the laws protecting domestic international property rights. They think weak laws will help them, so American firms and inventors find themselves confronting an almost hopeless situation internationally—inadequate laws defining the rights to be protected, inadequate lengths of term for patents, some governments requiring licensing as a condition to patent protection, inadequate anti-counterfeiting laws, and the absence of the consensus in the international community on the desirability of enforcement. Clearly, things are not going to improve until the Third World is convinced of the need to protect intellectual property rights and is willing to enforce protection. Why is that important? It is a key element in the emerging issue of competitiveness in America. Seventy-five percent of American jobs are now vulnerable to foreign competition. If we are going to avoid mounting domestic pressures for protectionism, government in America is going to have to move aggressively to assure that America can compete in the emerging world economy.

So the question is, what can the United States, which has the most at stake, do to strengthen international protection of intellectual property rights. Conceptually, we must make competitiveness a key element in American foreign policy. What does that mean? It could mean these things as opening ideas.

- Withhold most favored nation status for any country that refuses to enact and enforce strong laws regarding intellectual property rights.
- Re-think the negotiating strategy for bilateral agreements.
- Give Third World nations a stake in intellectual property rights by providing technical and financial aid if they will pass strong laws protecting intellectual property rights.
- Resist compulsory licensing as a condition to patent protection.
- Expand the GAP agreement to prohibit counterfeiting and also to cover intellectual property rights.

These are only beginning ideas. What else can we do? One of the things that is happening in the House is the holding of joint hearings between the Foreign Affairs Committee and the Science and Technology Committee in which we are trying to look at this functionally for the first time in the Congress. It may be that as the issue emerges, the Congress is going to have to re-structure itself just so we can speak functionally about the problem.

The focus of the hearings is science and technology in foreign policy. When you think about the emerging issue of economic competitiveness in the same light as we now think about military competitiveness, you can see that we have tremendous muscle and that we can enforce an order on the world if we will only ourselves begin to think functionally about what our problem is.

What else can we do? I don't know the answer to that. I am not sure it has been thought through. I believe this hearing this morning and the symposium that will follow will provide us with an excellent starting point, and it is my pleasure to be a part of this.

Testimony from Dr. Donald R. Price
Vice President for Research
University of Florida

Biotechnology has become a major topic of discussion for the past several years among university research administrators. The issues surrounding this new research field are numerous and many are extremely important. The debate among and between the universities and private industry has been rather fierce at times, but nearly always healthy and constructive. I am of the opinion that the differences and more controversial problems are diminishing. A standard practice, or norm, is emerging that better defines the issues and agreement is thus more easily attained.

The issues I am referring to are: (1) ownership rights to intellectual property generated from a research grant or contract from an industry to university researchers; (2) publication of research results without undue delays and restrictions; (3) ownership rights to data; (4) confidentiality requirements between universities and industry; and (5) patent licensing agreements to commercialize results of university research.

In no technological field have the issues been more controversial than in biotechnology. This is probably to be expected given the interest and economic potential of new discoveries. There are numerous new start-up companies all competing for survival. Because they are new and small, they need the research base and support available at the universities. One or two timely discoveries can make a company successful. Therefore, the new companies are quite aggressive about control of patents and licensing arrangements.

Let me proceed with a more detailed illustration of one issue that is often at the forefront of disputes between industry and universities. Many, in fact most, corporate-level officers in a company hold to the position that if the industry provides funding for a research project on a university campus, then any discoveries (patents) that result from the research should belong to the sponsor company. The universities take the position that if a faculty researcher makes such a discovery, it belongs to the university. The universities have very large investments in research faculty in the form of laboratory space and equipment, libraries, graduate students, technicians, etc. A faculty member usually develops expertise in a field over a period of several years and it would be unfair for a private industry to come along and reap all the benefits of that investment with one small research grant or contract.

The university should, for this reason and others, retain ownership and control of patents. As a public institution, we are obligated to get the research from the lab out to benefit the public, if we can. We have an arrangement to share any income that might result from a patent directly with the inventors. Our policy is to share 50 percent of the first $100,000, 40 percent of the next $100,000, and 30 percent of all income over $200,000 directly with inventors.

Now, to be fair with the industry that sponsors the research, we provide in the research contract for the industry to have the first option to an exclusive license to any patent resulting from the sponsored research project. Usually when a potential sponsor fully understands our practice and the rationale behind it, we have had very few disputes.

I believe the cooperation between industry and universities is good and it is necessary to bring about the greatest benefits to the public. I hope the federal government will continue to encourage this cooperation by continuing the tax credits for industry support of research on university campuses. This is a good investment in the future and needed to keep us competitive with foreign countries.

Universities and industries can team up to help maintain our edge in science and technology among the strong foreign competition. Biotechnology will likely play a major role in this worldwide arena in the years ahead.

PART II
ADVANCES IN BIOTECHNOLOGY: BREAKTHROUGHS AND BOTTLENECKS

An Overview of Biotechnology

Philip H. Abelson
American Association for the Advancement of Science
Washington, D.C.

In this presentation, some of the progress in biotechnology will be evaluated. The use of DNA technology in producing medical therapeutic and diagnostic agents and the applications of monoclonal antibodies and tissue culture will be discussed. The agricultural importance of rhizobia and mycorrhyzae will be mentioned briefly.

DNA TECHNOLOGY

Therapeutic agents derived from the application of recombinant DNA techniques have received much attention. Eventually, they will undoubtedly play an important role in medicine. Today, however, only three have been approved by the U.S. Food and Drug Administration (FDA) for therapeutic use. They are human insulin, growth hormone, and alpha interferon. The histories of these pharmaceuticals provide perspective on time tables and problems in creating proteins for therapy.

INSULIN

For more than 50 years, insulin extracted from the pancreases of cattle or swine has been used in treatment of human diabetes. However, animal-derived insulin is not identical to the human product, and in some instances this may lead to unwanted reactions. In addition, supplies of beef and pork insulin are limited, and a future scarcity has been envisioned. Insulin is a relatively small and simple protein whose structure is well known. Isolation of the gene giving rise to it was comparatively easy, and human insulin was one of the first proteins to be produced by recombinant DNA procedures. *Escherichia coli* were used in the process; the initial creation of the bacterial product occurred in 1979. Insulin with a proper human composition was an attractive product, with every reason for speedy production, clinical testing, and FDA approval. Nevertheless, it took seven years for human insulin, named Humulin, to achieve a significant role in the management of diabetes.

The full story of what happened has not been detailed. However, rumors are that problems were encountered in the use of *E. coli*. The insulin gene was well expressed, but the protein formed was retained within the bacterial cells. Complete separation of the product from the host of other *E. coli* proteins was absolutely essential to avoid pyrogenic activity, but this was not easy. The initial sales price of Humulin was much greater than that of animal insulin. Apparently, the production problems have been overcome, possibly by using another organism such as yeast as an expression system. In any event, Humulin is now broadly accepted. At least half of new diabetic patients in the United States are being treated with it. Humulin now sells for less than animal insulin.

In summary, experience with insulin demonstrates that important therapeutic agents can be produced by recombinant DNA technology. Costs can be competitive. The time required to go from laboratory to widespread use can be many years.

GROWTH HORMONE

The second therapeutic recombinant DNA drug to be approved by the FDA, a growth hormone for humans, also has a long history and favorable circumstances for its acceptance. Genentech first produced its version of a human growth hormone in 1979. Clinical trials began in 1981. At that time, the human growth hormone being used for therapy was derived by extraction of the pituitaries of human cadavers. The extracts were pooled before final processing. Ultimately, concern was expressed that the product contained harmful viruses, and in early 1985, the FDA cancelled approval of the human-derived growth hormone. This obviously had an expediting effect on the approval of the Genentech product which was granted in October 1985. Some 4,000 children who would otherwise be destined to be adult dwarfs are now being treated. It is estimated that ultimately a total of the order of 10,000 to 15,000 children will receive the drug. Costs are high. They range between $4,000 and $6,000 per patient per year.

Patients treated with the Genentech drug have responded well. However, the protein is not identical to the human hormone. Other small companies have created the human variety by recombinant DNA technology, and their products are now being tested clinically. Genentech has found it prudent to create its own new comparable product and is subjecting it to clinical trials. Clinical trials are expensive. Costs quoted range from $25 million to $60 million.

Following clinical trials, applications for approval must be submitted to the FDA. If some of the companies are not successful, they will suffer a crippling financial loss. If all the companies are successful, the market will be fragmented and probably competitive low prices will ensue. In that case, nobody wins. The possible out is a non-therapeutic and unethical use of the hormone in normal people in an attempt to increase their statural growth.

THERAPEUTIC AGENTS FOR CANCER

The third recombinant DNA therapeutic drug approved by the FDA is alpha interferon. The 1985 Biogen annual report stated that a Biogen-produced alpha interferon was licensed in Ireland. On June 4, 1986, the FDA announced approval of alpha interferon for use in treatment of hairy cell leukemia. This is a rare form of cancer that affects 2,000 to 3,000 Americans. In contrast, about five million Americans have a history of cancer. Treatment of the hairy cell leukemia with alpha interferon has been quite successful. Remission of the disease has been noted in 75 to 90 percent of the more than 2,000 patients tested. But comparable effects have not been seen in the treatment of major forms of cancer.

A substantial number of recombinant DNA therapeutic drugs are currently undergoing clinical trials. Besides alpha interferon, these include beta and gamma interferons, interleukin-2, tissue necrotic factor, colony-stimulating factors, tissue-type plasminogen activator, and a number of others.

Cancer is such a dreaded disease that developments promising therapy for it draw front-page attention. Recombinant DNA technology has repeatedly received great publicity and the consequent interest of financial people. The first big wave of financial support for the emerging biotechnology companies was largely based on the thesis that interferons would cure cancer. A large number of companies, including Genentech, Cetus, and Biogen, cloned the genes and have produced interferons. A paucity of news about success of the ongoing clinical trials indicates that the interferons have not lived up to the earlier high expectations. Rumor has it that there have been serious side effects. One way of gauging what has happened, or is happening, is to examine annual reports of public companies such as Biogen, Cetus, and Genentech. The reports of Cetus and Genentech contain considerable detail. Genentech mentions studies in which a combination of alpha and gamma interferons is employed. It also mentions the use of interferons combined with tumor necrosis factor. Another clinical study involves interferons plus conventional chemotherapeutic agents. The Cetus annual report for 1985 mentions clinical studies being performed on beta interferon. It states that the product has been tolerated well by patients and several responses were observed during the Phase I testing, even though trials were designed to test safety and not efficacy. The Cetus report put more emphasis on interleukin-2 as a cancer therapeutic agent. It said that interleukin-2 is tolerated well by patients and that early responses have been seen with Kaposi's sarcoma, chronic leukemia, renal cell carcinoma, colon carcinoma, and others in Phase I studies. A clinical study at the National Institutes of Health (Rosenberg, 1985) involving 25 patients has testified to efficacy of interleukin-2. Cetus, however, is hedging its bets by developing a number of other therapeutic agents for cancer, including tumor necrosis factor.

From the foregoing, it is evident that interferons alone are not magical silver bullets for cancer. However, it is likely that recombinant DNA technology will produce additional useful therapeutic agents and that combinations of them may prove fairly efficacious in some, but not all, forms of cancer.

TISSUE-TYPE PLASMINOGEN ACTIVATOR

From the standpoint of successful treatment of patients, tissue-type plasminogen activator holds promise of becoming one of the most important products of recombinant DNA technology. Heart attack is the leading cause of death in the United States. An estimated 1.5 million Americans will experience a heart attack this year, and 550,000 of them will die. To function properly, heart muscle requires an adequate supply of blood. When a clot forms in a coronary artery and blocks the flow of blood, a heart attack may follow. If blood flow can be restored within an hour or so, irreversible tissue damage can be stopped. Tissue-type plasminogen activator is a potent natural clot-dissolving agent which is present in small amounts in the circulatory system and acts to initiate normal clot-dissolving activities. However, when a massive clot forms, the body does not provide a sufficient amount of the enzyme to dissolve the clot. Tissue-type plasminogen activator is a protein of molecular weight 60,000. At least ten companies have cloned the gene for it, and the product is undergoing extensive clinical trials (Crawford, 1986). The product is proving superior to an existing agent, a streptokinase. Prospects are said to be excellent for early approval of the product by the FDA. Approvals will likely come one at a time, with delays in between.

Some general remarks about recombinant DNA technology follow. First, once a protein or its gene has been isolated, techniques are broadly available for expressing the gene. The early dependence on *E. coli* has vanished. Individual companies have available as many as ten expression systems, including bacteria, yeast, molds, and mammalian tissue. A year ago, a postdoctoral student told me, "Anybody can clone a gene." His statement reflected the situation in some developed countries, but not in most countries. An important bottleneck is availability of restriction enzymes and other substances including tagged biochemicals. Another statement made to me was, "Genentech has cloned every gene that management thinks might be of possible interest." That is not to say that everything has been done, or even most of it. Rather, the statement is made to indicate that as opportunities and needs arise, the necessary technology will be quickly applied, provided a financial or other incentive is present.

PROTEIN ENGINEERING

For several years, the possibilities of changing the content of a gene have been widely recognized. Progress has been made in this, some of which has been reported in the literature (Villafranca *et al.*, 1983). The procedure thus far has been to modify a gene slightly so that when the gene is expressed, one of the amino acids has been replaced by another. Experiments have shown that resultant enzyme activity may be enhanced, unchanged, or decreased. Coincident with this, the folding of the protein and its shape may be changed. Substantial progress is being achieved in understanding the mechanisms, and top-notch crystallographers and extensive computer calculations are involved. Already practical applications are emerging, and many more can be expected. In its 1985 annual report, Cetus discloses production of what it calls muteins of natural proteins. Cetus has found that both the biological activity of the protein and the ease of production can be enhanced by changing some of the amino acids. In particular, an improved interleukin-2 was achieved by this procedure. Obviously, a new chemical form is patentable, and this could be an additional crucial advantage. In the annual report of Genencor, genetic engineering of subtilisin is mentioned. This protein, which is a protease, has been engineered and expressed in as many as 80 variants. An amino acid at one point in the molecule has been replaced by all of the other 19 amino acids. Genencor has been exploring the properties of subtilisin with a view to using it as a protease in detergents. They have found that by replacing amino acids in the natural enzyme, activity of the product can be enhanced. Other changes noted include tolerance of pH, temperature, and oxidizing conditions.

The combination of expertise permitting design of protein structures with enhanced performance and the advantages of patents guarantee extensive and successful applications of genetic engineering to create new non-natural proteins. When used *in vitro*, there should be only advantages. However, when applied as therapeutic agents, some may have antigenic properties.

VACCINES

A major world public health problem is the lack of vaccines to immunize large populations in the less developed countries. As a result, life expectancies are 10 years and more below those of the developed countries, and many people suffer debilitating diseases. In part, the lack of vaccines results from inability to purchase them. In part, the problem is due to the fact that needed vaccines have not been produced for important tropical diseases.

Several U.S. companies have conducted research and development aimed at producing hepatitis B vaccine. A recent announcement (Bialy, 1986) states that human trials of an antimalarial vaccine have begun at the Walter Reed Army Hospital.

However, commercial biotechnology companies can be expected to give a very low priority to producing vaccines under present-day circumstances. The financial incentives are, if anything, negative. Past experience is that virtually all vaccines can have serious side effects. Incidence of these may be only one in 100,000 or less. But present-day damage suits often result in awards of a million dollars or more. In consequence, many makers of conventional vaccines have stopped producing them. To induce competent companies in the United States to create and produce new vaccines will require mechanisms to eliminate threats of huge damage suits and to provide financial incentives.

Under these circumstances, the developing countries should join cooperatively to create and produce vaccines to meet their needs. Following production of these agents there would be practical problems to face, for example, avoiding spoilage where refrigeration is non-existent and sanitation is poor.

A recent development at the Wadsworth Center for Laboratories and Research, New York State Department of Health in Albany, could be very helpful (Perkus *et al.*, 1985). Smallpox was eliminated through global use of vaccinia virus. A dried preparation could be transported safely and administered under primitive conditions. The New York workers have found that other antigens can be expressed when vaccination is performed. Using gene splicing, they incorporated additional antigens into the vaccinia virus. One successful formulation includes antigens of hepatitis B virus, herpes simplex virus, and influenza virus hemagglutinin. The use of vaccinia virus is not completely without hazard. In one out of 300,000 instances, very serious side effects occur. The New York group has entered into collaboration with a French pharmaceutical company.

Prospects are good for development of many vaccines for veterinary applications. Regulatory and legal problems may be encountered when attenuated viruses are employed in vaccines. However, antigens produced by recombinant DNA techniques should encounter less resistance. Several years ago, *Science* awarded the Newcomb Cleveland prize for an outstanding article on successful development of a protein-based vaccine for one form of foot-and-mouth disease (Kleid *et al.*, 1981). Annual reports of the major biotechnical companies do not mention animal vaccines. However, a number of vaccines have been developed, and they are described by Weppelman (this volume). In addition, at the University of Florida, work is proceeding on the important tick-borne diseases anaplasmosis, babesiosis, and heartwater. Millions of cattle world-wide are at risk of these diseases. The University of Florida is in a unique position to conduct research on tropical diseases. It has an active center for tropical animal health, expertise in biotechnology, and collaborative links with the Caribbean and Africa.

DNA PROBES

An increasingly important application of DNA technology is its use in diagnostic probes. These have been employed for some time, but recent improvements in techniques have greatly increased their usefulness and sensitivity. The technique applied in the past has been to lyse cells and to absorb their DNA on a support (such as filter paper) in single-stranded form. Subsequently, the strands are exposed to radioactively-tagged DNA. If the two DNAs match, the radioactivity is retained. Otherwise, it is washed away. Such a technique can be used to detect the presence of DNA from any source, including viruses. The DNA-probe techniques are likely to prove of special value in detecting viral plant diseases.

Cetus has recently obtained patents on a variant of the DNA probe that appears to be highly sensitive and does not require the use of radioactive isotopes. It thus is applicable in physicians' offices and other places where radioactive isotopes are not available. Instead of radioactivity, biotin is incorporated in the DNA probe. Biotin changes color when it interacts with streptavidin. Cetus has also issued a press release mentioning a DNA amplification procedure which they say achieves million-fold amplification in target DNA. In consequence of these various developments, Cetus has said that it has an enormously sensitive procedure for detecting DNA.

MONOCLONAL ANTIBODIES

Monoclonal antibodies already have been widely applied as diagnostic aids and are beginning to have therapeutic applications.

DIAGNOSTIC AIDS

From the standpoint of improvement in the practice of medicine, monoclonal antibodies used as diagnostic aids have thus far had much more impact than has recombinant DNA. As of May, 1986, the FDA had approved 140 diagnostic kits. These kits are used for *in vitro* tests, and hence applicants need only demonstrate efficacy. The number of approvals has been growing exponentially. In early 1983, the total number was about 20. In late 1984, the number was 55. In part, the large figure results from development by a number of companies of tests for the same organism, for example, *Chlamydia*. Monoclonal antibodies are easily produced. Once the relevant hybridoma is created, it can be immortal. Samples of it injected into abdominal cavities of mice result in the production of fluids containing antibodies for as many as 10,000 tests from each mouse. In the future, the bottleneck will be marketing rather than production.

The two outstanding pioneers in commercialization of monoclonal antibodies were Hybritech and Genetic Systems. These two started out as small biotechnology companies. They succeeded in rapid development of diagnostic aids. A pregnancy test by Hybritech was rapidly accepted. The test detected pregnancy 10 days after conception. Genetic Systems (Nowinski *et al.*, 1983) earlier produced diagnostic tests for venereal disease that are quick and very useful. The two companies have developed many other tests. A measure of their success and of the value of their products is that both companies have become subsidiaries of major pharmaceutical firms. In each case, the consideration involved was about $300 million.

THERAPEUTIC POTENTIAL

Those engaged in monoclonal antibody research are looking to frontiers beyond diagnostic aids. Recently this has resulted in the FDA approval of Orthoclone OKT*3, produced by the Ortho Pharmaceutical Corporation. The drug has proved useful in treatment of short-term episodes of kidney transplant rejection. Research teams are also developing therapeutic antibodies for intractable bacterial infections and are seeking agents for diagnosis and treatment of cancers.

Some 60,000 patients in hospitals die each year as a result of opportunistic infections by organisms that do not respond well to antibiotics. These organisms, *Klebsiella*, *Pseudomonas*, and *E. coli*, produce potent endotoxins. These toxins can be neutralized by antibodies. However, mice cannot be used to produce them. To avoid antigenic effects, the therapeutic antibodies must be produced from human cell cultures. This is being done.

Monoclonal antibodies have been heralded as potentially important in the diagnosis and treatment of cancers. Tumor cells produce antigens on their surfaces. These can be used to create corresponding antibodies. When such antibodies are injected into animals or humans, they tend to bind to the tumor cells. This may interfere with the cells' metabolism. Clinical trials are in progress testing this approach. Another procedure is to tag the antibody with a radioactive isotope of yttrium. This combination can be employed to detect metastases of a cancer. Hopes have also been raised about using the radiation from the isotope as a treatment modality. Another approach that has been tried is to couple a toxin such as ricin to the antibody in the hope that the toxin will destroy the tumor cell. These hopes have been around for several years, but conspicuous success has not been evident. Apparently, the agent which includes ricin and antibody is not sufficiently specific in its binding, and it affects normal cells as well as tumors. A procedure is now being investigated by Cetus in which only the toxic part of the ricin is attached to the antibody. This approach may prove more effective and less toxic to normal cells.

PLANT BIOTECHNOLOGY

Technology has been applied to agriculture and the preparation and preservation of food for a long time. Great improvements in yield and quality have occurred. In many respects, in the short-term, the new technologies will enhance rather than supplant the old. For example, conventional methods of plant selection, breeding, and field testing will not be abandoned. In comparison to our knowledge about human physiology and biochemistry, we are ignorant about the plant world. We are especially lacking in knowledge about the role of the microbial life that exists in the soil surrounding the roots of plants. We talk about altering the biochemistry of plants, but until recently have known little of the complex interactions that control how genes are turned off or turned on.

TISSUE CULTURE

The big recent development in plant biology is tissue culture. It is permitting the cloning of selected and virus-free plants. It permits more rapid improvement of many plants than could otherwise be achieved. Originally only a few dicotyledonous plants could be propagated in tissue culture, but the number is increasing rapidly. Culture of monocotyledonous plants from single cells has not proceeded well, but Winston Brill of Agracetus believes that eventually all plants will be propagated through tissue culture. Such an achievement would open the road to large-scale enhancement of the capabilities of plants.

Tissue culture is a necessary step in manipulations designed to select or achieve an improved plant. By exposing tissue cultures to adverse conditions such as salinity or pesticides, one can select those specimens most capable of meeting the adverse condition.

GENETIC ENGINEERING OF PLANTS

Tissue culture also is involved in the introduction of DNA into cells to change their genetic inheritance. Calgene has succeeded in introducing a gene that gives tobacco some resistance to the herbicide glyphosate. Glyphosate, also known under the trade name "Round-up," is a successful product of Monsanto. The chemical blocks a pathway in plants in the synthesis of aromatic amino acids. The gene that Calgene has introduced enables the plant containing it to continue synthesis of the aromatic acid in the presence of Round-up.

In the experiments described to date, the resistance to Round-up is somewhat limited (Marx, 1985). An additional enhancement of resistance would be necessary to achieve an excellent plant that could thrive in the presence of heavy applications of Round-up.

A group at Monsanto has engineered glyphosate-resistant petunia plants by inducing them to make 20 to 40 times the EPSP enzyme that is the target of glyphosate (Marx, 1985). This was done by attaching petunia EPSP synthase gene to a viral-regulating sequence that is a promoter of gene expression and then transferring the hybrid gene into the plants.

Recently, I learned of another instance of a successful introduction of a gene into a plant. My informant was Professor Lawrence Bogorad of Harvard University. He often collaborates with Belgian investigators and was in Belgium recently to see results of a trial of incorporation of a toxin gene into tobacco. The gene was derived from *Bacillus thuringiensis*, and a toxin elaborated by this gene is known to be highly toxic to caterpillars while not affecting other forms of life. The Belgian scientists had not only incorporated the gene into the genome of the plant, but they had managed to incorporate it in such a way that the gene was highly expressed. In perhaps an overstatement, but a colorful one, Bogorad said, "The caterpillars took one bite out of the plant and fell over dead. The control plants nearby were reduced to shreds."

The superior expression of the *B. thuringiensis* gene in tobacco contrasts with the situation that usually occurs in humans. In all our cells we have the same DNA, yet only a fraction of the genes are turned on, and those that are turned on change during development. In one case of the early work on recombinant DNA, a great amount of information was available about *E. coli*. It was possible to use that information to achieve enormous expression of a foreign gene by including it in a plasmid that carried functions essential to the growth of *E. coli*. In experiments with protoplasts, only a small fraction incorporate DNA from the medium into their genome and that DNA may or may not be expressed. However, investigators are learning how better to incorporate genes and to get them better expressed.

In any event, the simple culture of plant cells is having and will have major world-wide benefits. The technique is simple, the chemicals needed are readily available, and scientists in the developing countries can obtain and use them. Each terrain in each country presents a different circumstance, and the varieties of plants that can be selected and grown by tissue culture are innumerable. Active programs in tissue culture are being carried out in many developing countries, including Brazil, Colombia, Costa Rica, Cuba, Mexico, China, India, and Thailand.

RHIZOBIA AND MYCORRHYZAE

Plant biologists have long been aware of the role of rhizobia in nitrogen fixation and of beneficial effects of mycorrhyzae in capturing phosphate. Recent work has led to more systematic and effective use of these microorganisms. The results are particularly applicable to developing countries that are chronically short of hard currencies to pay for fertilizers. Rhizobia naturally present in a soil are not necessarily optimum for the legume that has been planted. On a recent visit to Brazil, I was told of a problem encountered when farmers attempted to grow soybeans in a region of the Cerrado. Nitrogen fixation was poor. The problem was traced to an *Actinomyces* that secreted an antibiotic injurious to the rhizobia. Through selection, a variety of rhizobia was obtained that is resistant to the antibiotic, and yields of soybeans improved. Considerable research is being conducted in other developing countries to discover improved varieties of rhizobia. Examples are India and Kenya. In Kenya, the effort has proceeded to the point where 10,000 farmers have been provided with rhizobium inoculants.

Mycorrhyzae associated with plants can have very important beneficial roles. Benefits can include enhanced efficiency of uptake of phosphate, drought tolerance, broader pH tolerance, and resistance to certain pathogens. These fungi attach to plant roots and send out hyphae that increase by, for example, a factor of ten the volume of soil tapped by the plant. Many woody plants have ectomycorrhyzal fungi as their symbionts. Techniques are available to grow the mycelia of these fungi in culture, and they are commercially available in the United States. Millions of pine seedlings are already being treated. Benefits include enhanced performance of seedlings in the nursery and better survival and growth in the field.

The symbionts of most food crops and some trees invade the roots, resulting in combinations called vesicular-arbuscular mycorrhyzae (VAM). Until just recently, inocula were not commercially available. However, Native Plants, Inc. (NPI) of Salt Lake City, Utah, has announced that it can provide inocula. The NPI product is in the form of spores. Recently I visited the laboratory of a subsidiary of NPI called Bio-Planta located in Campinas, Brazil. There I saw a demonstration of the efficacy of VAM in promoting the growth of lemon tree seedlings. The controls were only about half as tall as the treated specimens, and the controls did not look healthy. The scientist who showed me the seedlings said that he had tested 80 different isolates of VAM. Of these, six performed very well. The remainder were not much better than controls. Evidently much is to be gained in the health of plants and their growth by application of rhizobia and mycorrhyzae. In addition, the optimum symbiont for a given crop is probably a function of local conditions. Ultimately, it may be possible to engineer better symbionts, but it is also desirable to understand better the treasures that nature has already provided.

PERSPECTIVE AND CONCLUSIONS

Much of the future of biotechnology will be shaped by legal and regulatory considerations. In turn, the regulatory environment will be shaped by the tides of public opinion. It is probable that when there is a clear-cut obvious benefit, such as cure for a disease, the public will applaud. However, the public is easily made apprehensive, and proposals to release engineered microorganisms into the environment are likely to encounter tough going.

Probably the biggest factor shaping the future of biotechnology will be the patent situation. An enormous number of patents have been applied for and doing so has become highly fashionable. The 1984 report of Genentech stated that they had received 100 patents and had 2,000 applications pending. The 1985 report of Cetus mentioned 1,000 applications. Cetus press releases indicate that they have been successful in quite a few applications. Annual reports of other companies mention patents. It is likely that in coming years a considerable troop of lawyers will dine well on biotechnology.

Tremendous progress is being made in the application of fundamental knowledge. But because of the need for caution when preparing and administering therapeutics, the time required will be long for full demonstration of the power of recombinant DNA technology. Gene engineering is here, and it holds great promise. Monoclonal antibodies are already improving the practice of medicine. Tissue culture has been sufficiently successful that there is no doubt that it will have a great impact on world agriculture. The study and use of soil microorganisms will likewise make a major contribution. Those engaged in biotechnology are active in an area where great things have happened and will continue to happen.

REFERENCES

Bialy, H. (1986). Cloned malaria vaccine enters the clinic. *Biotechnology* 4: 384.

Crawford, M. (1986). Biotech market changing rapidly. *Science* 231: 12-14.

Kleid, D.G., Yansura, D., Small, B., Dowbenko, D., Moore, D.M., Grubman, M.J., McKercher, P.D., Morgan, D.O., Robertson, B.H., and Bachrach, H.L. (1981). Cloned viral protein vaccine for foot-and-mouth disease: Responses in cattle and swine. *Science* 214:1125-1129.

Marx, J.L. (1985). Plant gene transfer becomes a fertile field. *Science* 230: 1148-1150.

Nowinski, R.C., Tam, M.R., Goldstein, L.S., Kuo, C.C., Corey, L., Stamm, W.E., Handsfield, H.H., Knapp, J.S., and Holmes, K.K. (1983). Monoclonal antibodies for diagnosis of infectious diseases in humans. *Science* 219: 637-644.

Perkus, M.E., Piccini, A., Lipinskas, B.R., and Paoletti, E. (1985). Recombinant virus: Immunization against multiple pathogens. *Science* 229: 981-984.

Rosenberg, S. (1985). Observations on the systemic administration of the autologous lymphokine-activated killer cells and recombinant interleukin-2 to patients with metastatic cancer. *New England J. Med.* 313: 1485-1492.

Villafranca, J.E., Howell, E.E., Voet, D.H., Strobel, M.S., Ogden, R.C., Abelson, J.N., and Kraut, J. (1983). Directed mutagenesis of dihydrofolate reductase. *Science* 222: 782-788.

Impacts of Biotechnology on Agriculture: Plants

Peter R. Day
Plant Breeding Institute
Cambridge, England

BACKGROUND

This paper is concerned with plant molecular biology and agricultural crop plants. Although the first topic is too young to have had a major impact on the second, it is such a vigorous and rapidly-developing science that many people expect that this will occur soon. Depending on the nature of their interests, they want to be sure not to miss new opportunities to benefit from applications, and are anxious to monitor progress in the field and even influence the directions this may take. Even the most sober enthusiasts, and I count myself among them, recognize that the impact of molecular biology on our understanding of genetics and development could well have profound implications for those whose job it is to shape new materials for food and fibre. The early pipe dreams of nitrogen-fixing cereals, and entirely new ranges of crops and farm animals, have been put on the shelf for the time being, as many practicing scientists grapple with the realities.

During the years since the Second World War, the development of agronomic crops has made tremendous strides. In the developed world nearly all of the major crops, except perhaps for soybean, have shown a steady increase in yield which, in the last few years, has helped to generate, in North America and the European Community, embarrassing surpluses. There is no shortage of food in the developed world, and the difficulties of poorer nations are often seen as primarily due to the problems of transporting grain to where it is needed and of paying for it. Speaking as a plant breeder, there are two major challenges facing us at the end of the eighties. These are, to provide the crop varieties needed to supply food and raw materials for industry in the developed world, and to assist the less developed nations to become self-sufficient in agriculture by increasing productivity without increasing their dependence on chemical inputs. No matter which part of the world he works in, the breeder has to attend to three requirements: to increase yield, to improve the quality of the harvested crop, and to reduce production costs. This last includes resistance to pests and diseases, and environmental stress, relieving the need for applied pesticides and increasing the productivity of land areas and climatic zones not suited to growth and development of presently available crop varieties.

Breeding crop plants is a very successful operation that relies on good organization and management to handle large numbers of plants, sustained hard work, and the exercise of judgement born of experience. Successful breeders will use every trick and method they can find if they will help them to carry out their tasks more efficiently (Day, 1985). They are generally too busy to write many research reports or to spend time theorizing. If biotechnology can help them, they will be the first to want it, but they are understandably skeptical of the claims made by many of its exponents. It is important to be aware of this gap. It stems from the nature of the day-to-day work and interests of breeders and molecular biologists. The two disciplines use different terms. They involve work at either end of the spectrum extending from very large plant populations to cells, organelles, and molecules. A constructive dialogue does not necessarily occur automatically. Biotechnology provides a method of painstakingly disassembling the individual component parts of an organism and putting them together again. Anyone who has ever assembled a radio or similar electrical circuit with the expectation that it will work will know that it is necessary to have a circuit diagram, showing the components needed, how they should be laid out, and how they are wired together. Although the analogy quickly breaks down, since a living organism is an assembly of very complicated self-replicating circuits that have taken millions of years to evolve, plant molecular biology is still very much at the stage of elementary tinkering. This consists of taking things apart to see if and how they work when put back together and inserted, more or less at random, into a functioning organism such as a bacterium or yeast. All molecular biologists recognize that a very great deal more has to be done to improve our understanding of how plants work before we can be in a position to introduce the major changes that will no doubt be possible sometime in the future.

CURRENT ACHIEVEMENTS IN PLANT MOLECULAR BIOLOGY

This is neither the time nor the place for a comprehensive review of progress; however, we note that for crop plants like tobacco, potato, tomato, sunflower and oilseed rape (canola), the introduction of a foreign gene by transformation, using vectors based on the crowngall bacterium,

Agrobacterium tumefaciens, is now almost commonplace. All of these dicotyledons are natural hosts for *Agrobacterium*. The frequency of transformation, and the stability of transformants in subsequent generations, are both sufficiently high to be potentially useful to plant breeders. But important problems remain. As I shall mention later, there is no control at present over the sites of integration of introduced DNA or of the numbers of copies which become integrated. This means that each transformed plant is potentially unique in regards to its fitness and field performance.

For cereals and many other monocotyledons there are still major difficulties in effecting transformation. Until this can be done routinely, we will not be able to exploit the benefits of gene transfer in wheat, barley, maize, rice, and sorghum. Although it is possible to introduce DNA to protoplasts, using the method of electroporation, regeneration of plants from cereal protoplasts is proving to be extremely difficult technically. Microinjection of fertilized wheat ovules appears to be a promising method, since the tissue is programmed to develop into a seedling, but work to date has been unsuccessful. Lörz (personal communication) recently claimed to have introduced a bacterial gene for antibiotic resistance into rye by injecting a DNA preparation into shoot meristems. However, this, like earlier claims of Soyfer *et al.* (1976) for transformation of barley, needs confirmation.

Sidestepping the issue of cereal transformation, we may ask what valuable single genes are available for introduction. If the gene product can be identified, there are several methods for recovering cDNAs prepared from the messenger RNA which specifies the gene product, which can then be used to probe for the gene itself in a DNA library of the organism in a bacterium. Another method, called transposon tagging, depends on Barbara McClintock's discovery of jumping genes. Namely, that if you can identify a mutation in a character that is of interest, that is, the result of the gene controlling the character being inactivated by the insertion of a small DNA sequence called a transposon, then it is possible, by detecting the base sequence of the transposon DNA, to identify the gene by a series of probing steps. As a result of using these and other methods, a steadily-increasing number of plant genes have been identified and characterized. Of particular interest in this list are genes involved in the determination of self-incompatibility in *Brassica oleracea* (Nasrallah *et al.*, 1985) and *Nicotiana alata* (Anderson *et al.*, 1986). The structure of these genes should help to unravel the mystery of plant cell recognition systems that not only regulate pollen-tube growth in the style, but plant interactions with parasites.

Self-incompatibility is being used very effectively to prevent self-fertilization for the commercial production of hybrid seed of kale, another brassica. The work of Nasrallah *et al.* (1985) offers the promise of probes to detect the incompatibility genotypes of seedlings and the eventual possibility of directed changes by transformation in other important brassicas such as oilseed rape, where the potential for hybrid seed production is very much greater. Genes for herbicide resistance are likely to have early practical applications since they allow the plant breeder the opportunity to confer resistance to useful herbicides to which his crop is sensitive. This involves a role reversal, since until now the herbicide chemist has looked for chemicals that exploit the differential sensitivities of the crop he wishes to protect, and the weeds he wants to kill. There are also encouraging signs that we are beginning to understand the mechanism whereby genes are controlled during development, although there is still much to be found out about the nature of the signals which turn the controls on and off. An example of this is the promoter sequence controlling the element attached to the storage protein of beans, *Phaseolus vulgaris* (Sengupta-Gopalan *et al.*, 1985). When this gene and its promoter were introduced into tobacco, and transformed plants flowered and set seed, the introduced gene was only expressed at the time of seed formation. Even though the protein it formed was inappropriate for tobacco, the promoter attached to the gene evidently responded to signals that mark the onset of seed development which, it would seem, beans and tobacco have in common. There is considerable interest in introducing genetic information from microorganisms into higher plants. One example is the gene for a toxic polypeptide, the delta endotoxin of *Bacillus thuringiensis*, which has now been introduced into tobacco plants, conferring resistance to tobacco horn worm and other caterpillars. An alternative use for the same gene has been its transfer to a strain of *Pseudomonas fluorescens* isolated from corn roots (See Watrud *et al.* in Halvorsen *et al.*, 1985). Applied as a dressing to corn seed, the engineered bacterium colonizes the plant root surface and, in greenhouse tests, provides protection against the larvae of corn ear worm in the soil. Some work with virus genes has shown also that if a cloned cDNA of the gene specifying the coat protein of the tobacco mosaic virus particle is introduced into tobacco, it may confer resistance either by delaying disease development or by preventing virus infection by an as yet undetermined mechanism (Abel *et al.*, 1986). Many plant viruses have single-stranded RNA genomes. There is interest in transforming plants with genes which are transcribed to produce an antisense single-strand RNA complementary to the viral genome. Here the idea is that the two complementary RNAs, one of viral origin and the other now of plant origin, will interact to prevent transcription and multiplication of the viral genome (Palukaitis and Zaitlin, 1984, see also Baulcombe, 1986).

One important use of plant molecular biology in plant breeding is the production of specific probes, which can be used to detect moderately large blocks of genetic information in the breeder's plants. One of the best-known examples is the dot blot or sap spot test for detecting the presence of viroids and viruses in plants. The reagent used is a cDNA prepared to part of the RNA genome of the pathogen. This is then labelled in such a way that when it hybridizes to complementary RNA, and thus becomes bound to sap spots squeezed out and baked onto a cellulose nitrate membrane, these can be revealed when the excess unbound reagent has been removed by washing. The same method can be used to discriminate among different cytoplasmic male sterile lines of corn, which have characteristic differences in their mitochondrial DNA (Flavell *et al.*, 1983) to identify blocks of terminal heterochromatin in rye lines used as parents for producing the wheat/rye hybrid triticale and for detecting a chromosome arm from rye, present in a number of European wheat varieties, which has a deleterious effect on baking quality (Hutchinson *et al.*, 1985).

An interesting new development with great promise makes use of the fact that when DNA is extracted from a plant and treated with a restriction enzyme, it is cut into small pieces of characteristic size. If the restricted DNA is separated by electrophoresis on an agarose gel, and stained, it is usually revealed as a smear composed of many thousands of fragments of different length, which have migrated to different distances in the gel. However, by using a labelled DNA probe, for example, prepared by cloning a cDNA prepared from the plant's messenger RNA, or a fragment of DNA present in low copy number, and treating the gel so that only those fragments that bind to the probe are revealed (this is called a Southern blot), a much simpler banding pattern emerges. When different individuals within a population are compared, it is frequently possible to show that there are significant differences in the banding patterns. This is because the sizes of the individual fragments produced by the restriction enzyme vary, perhaps as the result of the deletion or addition of bases. These differences, known as restriction fragment length polymorphisms (RFLPs), can be used as markers in order to identify those parts of chromosome arms from which they are derived. RFLPs detected in the DNA from blood or semen samples provide a rigorous method of proving identity in forensic science or of familial relationship (Jeffreys *et al.*, 1985). For plant breeding, RFLPs show promise of providing a convenient and ready-made system of markers to develop linkage maps for any crop plant without the need for extensive genetic studies such as those carried out in corn and tomato (Burr *et al.*, 1983; Bernatzky and Tanksley, 1986). Thus, once characters that are of interest to breeders have been associated with RFLP markers, it should be possible to select only those progeny in segregating families which have the maximal number of characters the breeder seeks. This could reduce the number of backcrosses needed to introduce new characters from an alien wild species and also reduce the number of generations needed in pedigree selection programmes. Since the markers are co-dominant they may also be used to identify true breeding homozygotes. At the present time the method is laborious. It will be much more widely used when machines are developed to enable restriction digests, electrophoresis, Southern blotting, and gel recording to be carried out automatically.

Once a gene has been isolated and cloned, it is of course possible to modify its structure by altering the base sequence in places. By this means, it is possible to direct changes in the amino acid sequence of a peptide, with a view to changing the properties of the gene product. This may well be useful in breeding programmes concerned with improving quality. An example is breeding for improved baking quality in wheat, where wheat breeders seek to maximize endosperm protein glutenin sub-units with terminal sulphur-containing amino acids that, through disulphide-bonds, contribute to the cross-linking that occurs between these proteins in the development of dough required for bread-making (Day *et al.*, 1986). Figure 1 summarizes the principal interactions that are likely between conventional plant breeding and biotechnology.

CURRENT SCIENTIFIC ISSUES

Plant molecular biology is clearly a scientific discipline which has profound long-term implications for plant breeding. However, I have alluded to the difficulties in the way of achieving our expectations. One problem that tends to be neglected is the need to establish a productive dialogue between the individuals involved in the two activities. At the Plant Breeding Institute, we believe this is best handled by developing the technology that lies between them. At the moment this is largely concerned with the plant tissue culture needed to raise newly-transformed plants for evaluation and testing in collaboration with the plant breeders. The tissue culturist works with protoplasts, callus tissues, meristems, and embryos to select and regenerate potentially useful transformed plants and must therefore communicate effectively with the molecular biologists on whom he depends for genetic constructs. He is thus aware of the breeder's require-

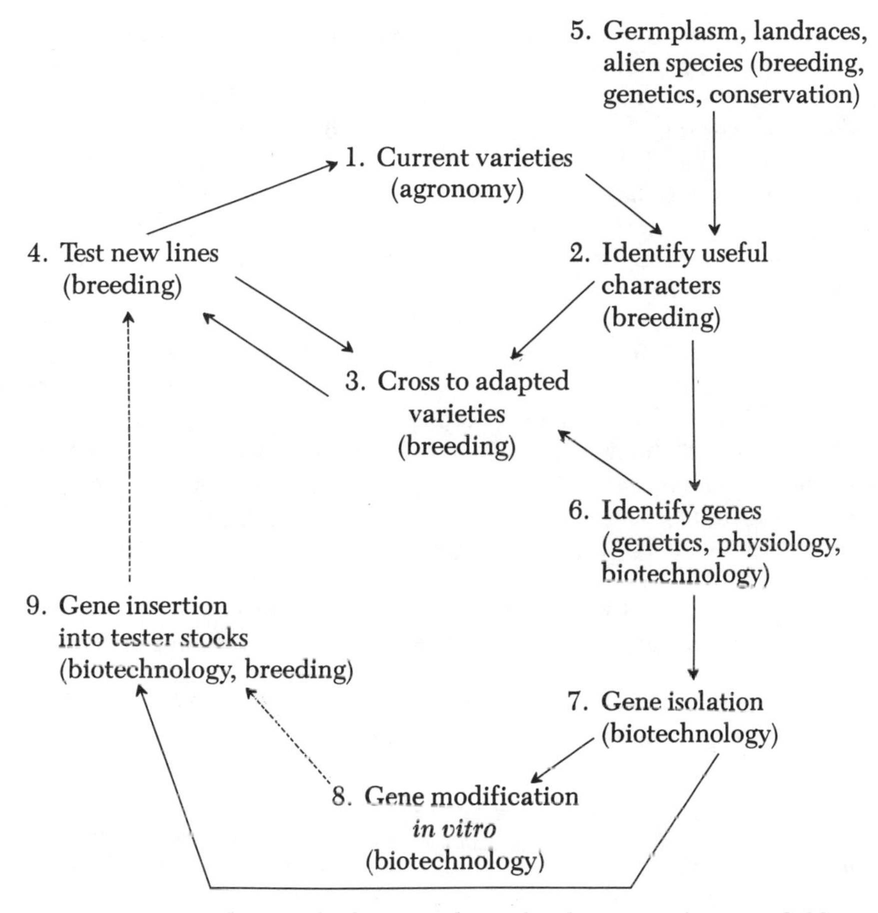

Figure 1. Breeding cycle showing the technologies used now (solid lines) and that are likely to be used during the next decade (dotted lines). Modified from Austin (1986).

ments and time-scale, but is also familiar with those of the molecular biologist. This work of technology transfer will be crucially important for success.

It is also clear that, for the foreseeable future, the application of biotechnology to shape new forms of crop plants will continue to depend very heavily on conventional plant breeding methods. Once new and interesting material begins to flow in their direction, plant breeders will evaluate it and incorporate some of it in their breeding programmes. There are many reasons why they will be cautious. For example, at the present time each product of transformation using the *Agrobacterium* system is potentially different from every other. This is because as yet there appears to be no control over the sites of integration of introduced DNA or the number of copies introduced. In a plant breeding context, therefore, the effect of integration site will have to be examined. Even if the molecular biologist has appeared to 'gild the lily' by adding one or more new characters to an established and widely-grown crop plant variety, the new line will need to be tested for stability of the introduced character(s) over several years and many trial sites, to make quite sure that transformation has not introduced an unanticipated defect which could render it valueless under certain circumstances. Plant breeders are used to uncovering the Achilles' heels in the most cherished products of their selection programmes.

Perhaps one of the most important scientific and social issues concerns framing sensible legislation to allow the introduction of products of genetic engineering into agriculture outside the containment of the laboratory and glasshouse, which is at present mandatory for all such materials (see Halvorsen *et al.*, 1985; Teich *et al.*, 1985). The difficulties encountered in arranging field tests of Ice-bacteria and of corn seeds treated with *Pseudomonas fluorescens* engineered to produce the delta endotoxin of *Bacillus thuringiensis kurstaki* are well known. The companies interested in commercializing these two microorganisms have spent millions of dollars, and thousands of man-hours, attempting to proceed through a thickening tangle of regulations and lawsuits. There can be no doubt that society must insist on safeguards to protect the environment. However, such safeguards have to be tempered by the knowledge that no-one can guarantee, with complete certainty, that any given released organism, whether or not genetically engineered, will have no adverse effects whatsoever. Answers must be provided to certain questions before field tests and environmental release are allowed. In my view, the requirement to produce, in an application for an experimental use permit, more than a thousand pages of evidence to show that an engineered organism is safe and harmless, would be absurd if it is to become the norm. Unless procedures are simplified, and sensible tests established, we will deny ourselves important benefits in a technology that may well greatly reduce our dependence on agrochemicals, which in the long run could be far more damaging both to us and the environment than products of genetic engineering.

SPECIFIC PRODUCTS AND TIME FRAMES

I may disappoint some readers by saying that I believe the time frames for biotechnology, at least during its early years, will probably be little different from those of conventional breeding. In our pedigree selection breeding programme for winter wheat at the Plant Breeding Institute, it takes from 10-12 years from the time a cross is made until the variety has reached the Recommended List and is available in quantity for farmers to buy and plant in their fields. By using the method of single-seed descent, it is possible to speed up the production of homozygous lines from F3 to F6, by cramming three generations into one year. However, since very little selection can be practiced under these conditions, the method at best only cuts two years off the total time. By using single-seed descent, our breeders could expect to have identified a candidate variety for submission to National List tests within five or six years. In Britain, National List tests take two years, and are followed by a further year of testing for Recommended List status. During this three-year period, seed stocks are built up to produce the 100 ton or so of basic seed needed for distribution to growers, who will raise certified seed for sale to farmers. Let us suppose that, as a result of genetic engineering, a plant breeder is presented with the equivalent of his best current variety that is either already in commerce or is about to go into the Recommended List, which now has a new engineered character, perhaps either herbicide or disease resistance. The first question would be, "Does the new form have sufficient merit to be a new variety in its own right, with all or most of the attributes of the original variety, plus the added character of resistance?" Even if the answer appears to be yes, the breeder will need two years of trials to satisfy himself on that point, and before the variety reaches a Recommended List, three more years of trials will be necessary. By that time, the original variety will already have been on the market for at least four or five years and may be outmoded by other developments from conventional breeding. New varieties that the engineered variety must compete with, might for example, have better quality or resistance to other diseases that are now more important. The breeder will have anticipated that this is likely to happen, and if he likes the material from the molecular biologist, will have introduced it into his own breeding programme to recombine the new character with the other features he is working towards. However, the product of genetic engineering may well take up to eight years to reach the farmer's field by following this route. It follows from all of this that it might be more profitable in the short-term to seek other ways of genetically engineering crop plants. For example, why not grow your crop in the field and introduce some benign systemic virus, which carries a package of genes that have the effect of promoting growth, altering the storage components in the endosperm of the seed, and which confer resistance to a range of fungal, bacterial, and insect pests. A step in this direction has already been taken by work to develop virus vectors. Brisson *et al.*, (1984) showed that a bacterial gene for drug resistance, introduced into the genome of cauliflower mosaic virus, was expressed in turnip plants systemically infected with the engineered virus. I won't go into the question of which of these two kinds of genetic engineering and applications would be the easiest to protect from the point of view of patenting. I don't think either of them would be particularly easy, although I suspect that engineered plants would more easily satisfy regulatory authorities than engineered viruses. Other opportunities for genetic engineering may well involve the production of arable crops which have entirely new properties from those we have at present. Not many years ago, we were intrigued by the success of chemists reshaping soybean protein into fibres that simulated the muscle fibres of meat. Perhaps one day we will be able to engineer tubers from a potato-like plant, which have a high content of both protein and fibre, to produce an entirely new textured form of vegetable to replace the meat in our diet. New developments of this kind may take longer but there will likely be no alternative but to wait for them.

CONCLUSIONS

The problem with any discussion on the impact of plant molecular biology on crop plant agriculture is that it soon degenerates into the realm of science fiction. I hope I have indicated for the general reader my enthusiasm for, and high expectations of, the very exciting work now going on in this field. Almost every week, in the journals that pour into libraries around the world, there are new discoveries and new vistas opening up which will give us cause to ask, "Is progress faster than we thought? Are the barriers to what is possible more easily surmountable than we thought?" I believe this will be the case, but I also believe that it would be foolish to neglect, in the meantime, the tried and tested technology on which we now depend for our food and fibre.

REFERENCES

Abel, P.P., Nelson, R.S., De, B., Hoffmann, N., Rogers, S.G., Fraley, R.T.

and Beachy, R.N. (1986). Delay of disease development in transgenic plants that express the tobacco mosaic virus coat protein gene. *Science* 232: 738-743.

Anderson, M.A., Cornish, E.C., Mau, S.L., Williams, E.G., Hoggart, R., Atkinson, A., Bonig, E., Grego, B., Simpson, R., Roche, P.J., Haley, J. D., Penschow, J.D., Niall, H.D., Tregear, G.W., Coghlan, J.P., Crawford, R.J. and Clarke, A.E. (1986). Cloning of cDNA for a stylar glycoprotein associated with expression of self-incompatibility in Nicotiana alata. *Nature* 321: 38-44.

Austin, R.B. (1986). *Molecular Biology and Crop Improvement.* Cambridge University Press, p. 114.

Burr, B., Evola, S.V. and Burr, F.A. (1983). The application of restriction fragment length polymorphism to plant breeding. In *Genetic Engineering* 5 (J. K. Setlow and A. Hollaender, eds), pp. 45-59. Plenum Press, New York.

Baulcombe, D.C. (1986). The use of recombinant DNA techniques in the production of virus resistant plants. In *Biotechnology and Crop Improvement and Protection* (P.R. Day, ed.), pp. 13-19. British Crop Protection Council, Thornton Heath.

Bernatzky, R. and Tanksley, S.D. (1986). Toward a saturated linkage map in tomato based on isozymes and random cDNA sequences. *Genetics* 112: 887-898.

Brisson, N., Paszkowski, J., Penswick, J.R., Gronenborn, B., Potrykus, I. and Hohn, T. (1984). Expression of a bacterial gene in plants by using a viral vector. *Nature* 310: 511-516.

Day, P.R. (1985). Crop improvement: breeding and genetic engineering. In *Technology in the 1990s: Agriculture and Food* (K. Blaxter and L. Fowden, eds.), pp. 47-54. Royal Society, London.

Day, P.R., Bingham, J., Payne, P.I. and Thompson, R.D. (1986). The way ahead: wheat breeding for quality improvement. In *Chemistry and Physics of Baking* (J.M.V. Blanshard, ed.), Royal Society of Chemistry, London (in press).

Flavell, R.B., Kemble, R.J., Gunn, R.E., Abbott, A. and Baulcombe, D. (1983). Applications of molecular biology in plant breeding: the detection of genetic variation and viral pathogens. In *Better Crops for Food, Ciba Foundation Symposium* 97, pp. 198-209. Pitman, London.

Halvorsen, H.O., Pramer, D. and Rogul, M. (1985). Engineered Organisms in the Environment: Scientific Issues. American Society for Microbiology, Washington.

Hutchinson, J., Abbott, A., O'Dell, M. and Flavell, R.B. (1985). A rapid screening technique for the detection of repeated DNA sequences in plant tissues. *Theoret. Appl. Genet.* 69: 329-333.

Jeffreys, A.J., Wilson, J. and Thein, S.L. (1985). Individual-specific 'fingerprints' of human DNA. *Nature* 316: 76-79.

Nasrallah, J.B., Kao, T.H., Goldberg, M.L. and Nasrallah, M.E. (1985). A cDNA clone encoding an S-locus specific glycoprotein from Brassica oleracea. *Nature* 318: 263-267.

Palukaitis, P. and Zaitlin, M. (1984). A model to explain the 'cross-protection' phenomenon shown by plant viruses and viroids. In *Plant Microbe Interactions* (T. Kosuge and E. Nester, eds.), pp. 420-429. Macmillan, New York.

Sengupta-Gopalan, C., Reichert, N.A., Barker, R.F., Hall, T.C. and Kemp, J.D. (1985). Developmentally regulated expression of the bean b-phaseolin gene in tobacco seed. *Proc. Nat. Acad. Sci.* USA 82: 3320-3324.

Soyfer, V.N., Kartel, N.A., Chehalin, N.M., Titov, Y.B., Ceincinis, K.K. and Turbin, N.V. (1976). Genetic modification of the waxy character in barley after an injection of wild-type exogenous DNA. Analysis of the second seed generation. *Mutation Research* 36: 303-310.

Teich, A.H., Levin, M.A. and Pace, J.H. (eds.) (1985). *Biotechnology and the Environment: Risk and Regulation* p. 201. Amer. Assoc. Adv. Science., Washington, D.C.

Impacts of Contemporary Biotechnology on Animal Science

Roger M. Weppelman
Merck, Sharp & Dohme Research Laboratories
Rahway, New Jersey

INTRODUCTION

The term "Biotechnology" is not easily defined. By the more inclusive definitions, animal science itself qualifies as one of the oldest (and most successful) of the biotechnologies. Rather than attempting to define "Biotechnology," I will simply state at the onset that the scope of this manuscript will be restricted to those aspects of contemporary animal science which have evolved from one or more of the following three papers in the scientific literature:

1. Chung and Cohen (1974). This paper describes the use of a restriction enzyme to create a functional genetic element bearing genes from two species of bacteria and thereby marks the beginning of genetic engineering.
2. Kohler and Milstein (1975). The authors fused a normal mouse spleen cell secreting a single type of antibody with an immortal mouse myeloma cell to produce the first hybridoma, which continued to secrete antibody and was immortal. This was the start of hybridoma technology and monoclonal antibodies.
3. Palmiter *et al.* (1982). These authors describe the creation of transgenic mice. Strictly speaking, this was not the first creation of transgenic mice, but because the gene used was beautifully engineered, because the technique was elegant and widely applicable, and because the results were truly dramatic, this paper marks the beginning of transgenic animals.

GENETIC ENGINEERING

Restriction enzymes are the basis for genetic engineering because of their unique ability to cleave double stranded DNA in such a way that the pieces can be readily rejoined (Maniatis *et al.*, 1982; Rodriquez and Tait, 1983). Hybrid genes can be engineered by cleaving two unrelated chromosomes with a restriction enzyme and then rejoining pieces of one chromosome with those of another. Insertion of a properly constructed hybrid gene into a suitable organism will enable the organism to produce the protein specified by the hybrid gene. If the protein is an enzyme, then the organism's metabolism might be altered to produce new metabolic products, or perhaps old products in better yields. In a conceptual sense, genetic engineering has the potential to yield the following classes of products: (1) recombinant genes; (2) organisms containing recombinant genes; (3) proteins made from recombinant genes; and (4) the products of enzymes produced from recombinant genes.

The product class comprised of recombinant genes will probably find application in correcting genetic defects in humans but is unlikely to be widely used by the animal industry. The second product class, organisms bearing recombinant genes, is discussed below under Vaccines and Transgenic Animals. A third possible product of this class is rumen microorganisms engineered to improve the efficiency of fermentation. The final class listed above, the products of enzymes specified by recombinant genes, could be extremely important since many products currently marketed for the animal industry are produced by microorganisms during fermentation. Among these are the coccidiostats and rumen additives monensin, salinomycin and lasalocid, the anthelmentic ivermectin, and numerous antibacterials, vitamins and essential amino acids. Insertion of the appropriate genes into the production microorganisms could lead to more efficient fermentations or to the production of improved derivatives of the original drug.

The type of potential product which has caused the most excitement todate is the class comprised of proteins synthesized from recombinant DNA. Before genetic engineering, most of these proteins were available in amounts which were adequate for only the smallest of laboratory experiments. These proteins are now available in amounts sufficient for large-scale studies of their activities and they have the potential to become available in the quantities needed to satisfy the needs of the world's animal industry. Three general types of proteins produced by recombinant DNA technology are discussed below.

HORMONES

The ability of bovine pituitary hormones to stimulate dairy cattle to produce more milk was suggested in 1937 by the observation that extracts prepared from bovine pituitaries enhanced milk yield (Asimov and Kouze,

1937). This effect was intensively studied in Britain as a way of increasing milk supply during World War II (Young, 1947). Subsequently, the active component within the pituitary extracts was shown to be bovine growth hormone (Brumby and Hancock, 1955). This observation remained only a curiosity until the gene was cloned and the hormone was produced in quantity by microorganisms. A recent publication by Bauman *et al.* (1985) demonstrated that methionyl bovine growth hormone, made by *E. coli*, increased milk production over a 188 day period by 23 to 41 percent, depending on the daily dose. Just as importantly, the efficiency with which feed was converted to milk improved up to a maximum of about 10 percent. These data indicate that bovine growth hormone could enable the dairy industry to satisfy the current demand for milk with an estimated 30 percent fewer cattle and at a considerably reduced cost in feed (*Animal Pharm*, November 14, 1985, p. 15).

It should be noted that the increases in milk production induced by growth hormone did not lead to net losses of body weight. Indeed the cattle gained weight while being treated with growth hormone and the gains were comparable to those of untreated cattle during the same period. This observation should put to rest concerns that growth hormone will strain dairy cattle by causing them to produce excessive milk (*Biotechnology News*, April 18, 1986, p. 1).

A very novel alternative use of bovine growth hormone to stimulate milk production was recently patented by Bauman and Sejrsen (U.S. Patent No. 4,521,409). These inventors discovered that treatment of dairy cattle during adolescence stimulates mammary development and increases milk output after their first calf. Whether treating with additional growth hormone during lactation will further enhance milk production remains to be determined (*Genetic Technology News*, July 1985, p.3).

The role of growth hormones in regulating growth has been recognized for years and for most species reductions in circulating growth hormone levels have been unambiguously associated with reduced growth. However, until very recently it was questionable whether giving additional growth hormone to "normally" growing animals would improve either their growth or utilization of feed. The answers for the important stock animals have been mixed. Increasing growth hormone levels in chickens three- to ten-fold did not affect growth rate (Souza *et al.*, 1984). When growth hormone purified from ovine pituitaries was given to growing wether lambs, growth was improved slightly but not significantly and efficiency was improved significantly by about five percent (Muir *et al.*, 1983). In a more recent experiment (Johnsson and Hart, 1985), treatment of growing female lambs with growth hormone from bovine pituitaries significantly stimulated both gain and efficiency by 22 and 12 percent, respectively. These authors noted that the hormone caused a 50 percent increase in fleece weight. Dramatic effects of growth hormones have also been reported in swine. When Baile *et al.* (1983) treated growing pigs with daily doses of recombinant human growth hormone, growth was stimulated up to 10 percent, but efficiency was not affected. Chung *et al.* (1985) reported that daily doses of growth hormone from porcine pituitaries not only stimulated growth by about 10 percent but also improved the efficiency of feed utilization by four percent. The only undesirable effect noted was an increase in the fat content in certain cuts of meat. Porcine growth hormone has been cloned and has been reported to be under development as a commercial product (*Animal Pharm*, November 29, 1985, p. 18).

Two problems must be solved before any of the hormones discussed above become an accepted tool of the animal industry. The first is that the cost of treatment must be a relatively small fraction of the economic benefit afforded by treatment. This is largely a matter of improving processes for production and purification of hormones and the problems are no different in principle from those which have been solved for every antibiotic marketed to-date. The second problem is more challenging. In all the experiments above, the hormones were given as daily injections for extended periods of time, a regimen which would probably not be acceptable to the animal industry. What is needed is an implantable depot which releases an appropriate quantity of hormone every day for extended periods. Several slow release technologies are available, among which are the osmotic pumps of the type pioneered by Alza Corporation (Palo Alto, CA), which release hormone at a rate depending on the osmotic pressure on the vessel containing the hormone. Also available are two types of bioerodable polymer. In one type, hormone is embedded in a matrix and slowly released as the matrix is hydrolyzed by the enzymes of the animal. In the second type, hormone is encapsulated by bioerodable polymer and released when the capsule ruptures. A condition approaching slow release can be achieved by using many small capsules which have walls of different thickness. In spite of the availability of these and doubtlessly other approaches, the development of a convenient method to assure the controlled release of active hormone for the extended time periods needed by the animal industry is not a trivial problem and appears to be the only remaining obstacle to the commercialization of the growth hormones.

VACCINES

Bachrach (1985) reviews more than 20 potential antiviral vaccines for animal health, all of which are based on protein antigens produced by recombinant DNA technology. He also describes an antiprotozoal vaccine against coccidiosis in poultry and several antibacterial vaccines. The majority of these vaccines would be impossible were it not for the ability of recombinant DNA technology to produce the antigens in the quantities required. In the remaining cases, traditional vaccines are either available or potentially available but vaccines based on recombinant DNA offer the promise of greater safety, enhanced efficacy, or reduced cost.

The vaccines described above are all of the "subunit type" which contain the antigen needed to induce immunity as their only active component. An alternative way of inducing immunity is to vaccinate with attenuated organisms, which are usually derived from the pathogen of interest and can induce immunity against it, but which lack its virulence. The Sabin polio vaccine is a classic example of a vaccine consisting of living organisms whose virulence has been attenuated. Recombinant DNA technology can also play important roles in the development of such living vaccines. Traditionally, attenuated organisms have been isolated by selecting avirulent mutants from a population of virulent parental organisms. Even when mutagens were employed to increase the frequency of attenuated mutants, the mutants were usually quite rare vis-a-vis the virulent parental organism and their isolation required a well designed selection and considerable persistence. Even after the attenuated mutants had been isolated, much effort was needed to demonstrate that the mutants did not revert to virulence, since a vaccine which could occasionally cause the disease is obviously unacceptable. Recombinant DNA technology allows one to introduce mutations of virtually any type into specific regions of the genome. Thus, it is fairly simple to place a deletion, which will not revert, into a gene known to be required by a particular pathogen for virulence. In addition to being faster and more convenient than the traditional approach, the recombinant approach offers an important safety advantage in that both the type of mutation and its location are known. By the traditional approach, neither the type nor location of the mutation conferring a virulence is known, nor is it known whether mutagenesis has introduced undesirable mutations elsewhere in the organism's genome.

The pseudorabies vaccine for swine, which has generated considerable controversy because it is the first vaccine containing engineered organisms to be approved by the USDA (*Chemical & Engineering News*, April 14, 1986, p. 4; April 21, 1986, p. 7; April 28, 1986, p. 18;*New York Times*,

April 13, 1986), provides an example of the use of recombinant DNA technology to attenuate a pathogen. In this particular case, the gene encoding an enzyme which allows the virus to attack nerve tissue was deleted (*Biotechnology News*, April 18, 1986, p. 1). Because the mutation responsible for attenuation is a deletion, reversion to virulence is virtually impossible.

INTERFERONS AND INTERLEUKINS

Viral interference, which refers to the relative resistance of a cell already infected with virus to infection by a second virus, was first noted in the 1930s (Hoskins, 1935; Findlay and MacCullum, 1937). The term "Interferon" was coined two decades later by Isaacs and Lindenmann (1957) who demonstrated that the resistance was due to soluble factors produced by infected cells. These factors were subsequently shown to be proteins and three general types of interferon have been identified in humans: alpha, beta and gamma. The three classes are chemically distinct and tend to be produced by different cell types (Friedman and Vogel, 1983): alpha interferon by leucocytes; beta interferon by fibroblasts; and gamma interferon by lymphocytes. Genes for all three have been cloned and, while there appear to be only single genes for beta and gamma interferon, there are a minimum of 14 genes for alpha interferon which vary 15 to 30 percent in amino acid sequence (Friedman and Vogel, 1983; Derynck, 1983). Whether these are alleles or "pseudogenes" (i.e., DNA sequences which are not expressed) is not clear.

The biological actions of the interferons are broadly similar. In addition to being antiviral, they inhibit replication of both normal and tumor cells (Brouty-Boyé, 1980). All three classes modulate the immune response in ways which are difficult to predict (Levy and Riley, 1983; Epstein and Epstein, 1983; DeMaeyer-Guignard and DeMaeyer, 1985). Whether the response is augmented or attenuated depends on the antigen, the immune response measured, and the treatment schedule (i.e., was interferon given before, after, or at the same time as antigen).

In view of the wide range of activities, it is not surprising that the interferons have rather serious side effects. According to Scott (1983), "It has become apparent that interferon itself is not innocuous; indeed it was never reasonable to consider that it would be." All interferons appear to be pyrogenic and induce in humans a collection of side effects which can best be described as "flu" (Scott, 1983). Growing children congenitally infected with virus stopped gaining weight when treated with alpha interferon and both alpha and gamma interferons have caused weight loss in cancer patients. Whether such severe side effects will also occur in stock animals remains to be seen.

The majority of animal health research on the interferons has been directed to bovine shipping fever. This flu-like disease of the respiratory tract, which appears to be caused by the stress associated with the transfer of feeder calves to feed lots, costs the U.S. cattle industry an estimated $300 to $700 million per year (Klausner, 1984). Shipping fever is also called bovine respiratory disease complex to emphasize that a rather large number of different pathogens, bacterial as well as viral, are commonly involved. The start of large trials to test the efficacy of human alpha interferon, bovine interferon, and an engineered consensus alpha interferon has been announced (Klausner, 1984; *Genetic Technology News*, April, 1985, p. 8). Publication of the results of these trials appears to be pending.

Interleukins 1 and 2, which are proteins that modulate the immune response, might also be efficacious against shipping fever and other diseases of stock animals. Interleukin 1, which is produced by macrophages and certain other types of cells, stimulates T lymphocytes to produce interleukin 2. Interleukin 2 in turn causes T lymphocytes to proliferate (Lachman and Maizel, 1980). Since T lymphocytes play several important roles in stimulating the responses to infection, both interleukins could be therapeutic for a variety of diseases. Both interleukins could also be useful adjuvants in vaccinations since they might increase the immune response if given simultaneously with antigen. The adjuvant effect of interleukin 2 has been shown in mice by Wood *et al.* (1983). A gene for murine interleukin 1 (Lomedico *et al.*, 1984) and two genes for human interleukin 1 (Mosley *et al.*, 1985) have been cloned. Bovine interleukin 2 (*Genetic Technology News*, April 1986, p. 3) and a modified human interleukin 2 (*Animal Pharm*, October 18, 1985, p. 10) have also been cloned.

It should be cautioned that interleukin 1 has many actions in addition to stimulating T lymphocytes to produce interleukin 2. Among these are pyrogenicity, stimulation of various inflammatory processes, and enhancement of muscle protein degradation (Gery and Lepe-Zuniga, 1984). Because of these side effects, interleukin 1, even if efficacious against various diseases, will probably find only limited application in animal health. Whether interleukin 2 shares these liabilities remains to be determined.

APPLICATION OF RESTRICTION FRAGMENT LENGTH POLYMORPHISM (RFLP) ANALYSIS TO THE BREEDING OF STOCK ANIMALS

The potential products of recombinant DNA technology discussed elsewhere in this paper either contain recombinant genes or are the direct products of recombinant genes. RFLP analysis will create no unique products. It has, however, the potential to accelerate conventional breeding programs and ultimately to lead to a full understanding of the genetics of stock animals.

Because the gene and gene products responsible for the traits valued in stock animals are not known, breeding has always been completely empirical. One cannot tell if a particular cross has been successful until the progeny are tested which in some cases can take considerable time. If the desired result is a dairy bull which will sire cows that produce superior quantities of milk, then five or more years might elapse before the success of the cross can be determined. RFLP analysis has the potential to shorten this time considerably. In the example of the dairy bull, one might be able to collect his cells by amniocentesis, subject his DNA to RFLP analysis, and thereby determine before his birth his potential as a sire of dairy cattle.

RFLP analysis (Botstein *et al.*, 1980) is based on the extremely specific endonuclease activity of restriction enzymes. Because of their specificity, restriction enzymes cleave DNA at a relatively small number of sites to yield fairly long fragments. The fragments can be separated electrophoretically by their size and then detected with homologous probes which have been cloned from a genetic library. For most regions of the chromosome, all members of a species will yield fragments of the same length. There are, however, chromosomal regions for which members of a species are polymorphic. Individuals which are homozygous for these regions will yield only short or only long fragments while heterozygous individuals will yield both short and long fragments. One can imagine that the "long" allele lacks a cleavage site present in the short allele or alternatively that the long allele has the same two cleavage sites but with an additional length of DNA inserted between them. Botstein *et al.* (1980) have estimated that 150 polymorphic sites, if appropriately located, would suffice to analyze the entire human genome.

The utility of the polymorphic sites for animal breeding is illustrated by returning to the example of the dairy bull. If both of the bull's parents are heterozygous ("short plus long") for a particular site, then the progeny will be either heterozygous "short plus long" like the parents, homozygous "long" or homozygous "short." The expected ratio in percents of the three types of progeny is 50:25:25 respectively. If one knew that superior milk production was associated with the short allele then RFLP analysis would permit one to select the 25 percent of the male progeny which are homozygous for the short allele and therefore likely to be homozygous for the genes for superior milk production.

The major obstacle to applying RFLP analysis to stock animal breeding is that one must know which polymorphic loci are associated with a particular trait. This involves the use of different restriction enzymes and probes to identify a large set of polymorphic loci. These loci must then be applied to individual animals whose pedigrees and performance traits are well-documented to develop correlations between the loci and performance. A substantial investment of effort would be needed but there appear to be no major obstacles to developing lists of polymorphisms associated with particular traits. The major advantage of this approach is that it requires no prior knowledge of the genes and gene products responsible for the traits and it is in fact a powerful tool for identifying those genes and their products. The power of RFLP analysis is proven by its success in partially mapping the human gene for Huntington's disease and in identifying carriers of the gene (Gusella *et al.*, 1983). These accomplishments are all the more remarkable in view of the fact that absolutely nothing is known about the primary genetic defect responsible for this neurodegenerative disease.

HYBRIDOMA TECHNOLOGY

Hybridoma technology is based on the fusion of immortal tumor cells with normal antibody producing cells to generate hybridomas which combine in single cells the properties of antibody production and immortality. Because hybridomas are immortal, they can produce antibody in quantities and for periods of time which are virtually unlimited. In contrast, traditional (or "polyclonal") antibodies can be obtained only in quantities limited by the size of the animal immunized and for time periods dictated by the animal's life span. Because all cells of a hybridoma are descendants of a single antibody producing cell, all antibodies produced by a particular hybridoma are "monoclonal" and are exact copies of each other: they share a common amino acid sequence and thus bind to the same site of the antigen with identical affinities. In contrast, polyclonal antibodies to a particular antigen are summations of many monoclonal responses and are thus heterogeneous. An important consequence of the homogeneity of monoclonal antibodies is that monoclonals can be treated as chemical reagents and readily linked to other molecules to increase their utility.

The initial impact of monoclonal antibodies on animal science is in diagnostics and Table 1 presents a partial listing of potential diagnostics, all based on monoclonal antibodies which have been described in the scientific literature during the past two years. This list is restricted to diagnostics for infections and does not include those for endocrine status, embryonic sex, or feed contaminants. In addition, *Animal Pharm Review* (January 3, 1986, p. 14) described 24 veterinary diagnostics which were introduced during 1985. The majority of these are based on monoclonal antibodies and are designed to diagnose either infections or reproductive status.

TABLE 1: Potential Diagnostics for Infectious Diseases, 1984-1986.

Organism (disease)	Reference
COMPANION ANIMALS	
Dirofilaria immitis (canine heart worm)	Weil *et al.* (1984)
Parvovirus (canine parvo infections)	Teramoto *et al.* (1984)
STOCK ANIMALS	
Mareks disease virus (Mareks disease of poultry)	Silva and Lee (1984)
Avian leucosis virus (poultry leucosis)	Boer and Osterhaus (1985 a,b)
Leptospira interrogans (bovine mastitis)	Stevens *et al.* (1985) Ainsworth *et al.* (1985)
Brucella abortus (spontaneous abortions of cattle)	Quinn *et al.* (1984)
Blue tongue and epizootic haemorrhagic disease viruses (bovine B.T. & E.H.D.)	Jochim and Jones (1984)
Bovine enteric cornavirus (calf scours)	Crouch *et al.* (1984)
Escherichia coli K99 (calf, lamb and piglet scours)	Holley *et al.* (1984) Mills and Tietze (1984) Morris *et al.* (1985)
Trichinella spiralis (swine trichinosis)	Gamble (1984)
All *Salmonella* (contaminant of many animal products, especially those from poultry)	Mattingly (1984)

A publication by Teramato *et al.* (1984) describing the development of a diagnostic for parvovirus in dog feces illustrates the utility of monoclonal antibodies. The authors selected two different monoclonal antibodies which bound to different regions of the virus hemagglutination protein so that a single antigen could simultaneously bind both types of antibody. One of the pair served as the "capture antibody" and was bound to the wells of microtiter plates. The second served as the "signal antibody" and was covalently linked to the enzyme horseradish peroxidase, which can be detected by its ability to turn a colorless substrate into an intensely colored product. The assay consisted of three steps. First, a fecal suspension was added to the wells containing the capture antibody, incubated to permit any hemagglutination protein present to bind, and then removed. In this step, the capture antibody is used to purify antigen from the fecal sample. Second, the signal antibody was then added and incubated to permit it to bind to hemagglutination protein which had been immobilized by the capture antibody in the first step. Unbound signal antibody was then removed. Third, substrate for horseradish peroxidase was added and, after a short incubation, color changes were noted. Those wells which turned color contained fecal samples which were positive for parvovirus. One can envision this assay supplied as a three component kit consisting of a solution containing signal antibody, a solution of substrate for horseradish peroxidase, and a microtiter plate whose wells had been pre-coated with capture antibody. Such a kit could be employed and interpreted by anyone who could read and follow instructions. Ultimately, kits of this type will place extremely sophisticated diagnostic tools into the hands of the average veterinary practitioner and stockman. This will result in a reduction in the time which lapses until appropriate therapy is initiated and should, as a consequence, decrease losses due to mortality and morbidity. Prompt and accurate diagnosis should also cause decreases in the inappropriate and prophylactic uses of drugs.

Monoclonal antibodies might also find application as vaccines and other types of biological effectors. The topography of a binding site of an antibody is complimentary to the topography of the antigenic site to which it binds. It follows then that the binding site of a second antibody, which is directed against the binding site of the first antibody, will topographically resemble the antigen to which the first antibody binds. Second antibodies of this type are called "anti-idiotypic" or "antiparatopic" antibodies. This method is similar to the lost-wax process used by sculpturers. A sculpted wax object, analogous to the antigen, is used to form a ceramic mold, analogous to the first antibody, which in turn is used to form replicas of the object, which are analogous to the anti-idiotypic antibody. Bachrach

(1985) reviews several anti-idiotypic antibodies which have shown potential as vaccines.

At least in one case, monoclonal antibodies have been shown to be effective prophylactic agents (Sherman *et al.*, 1983). Certain strains of *E. coli* can colonize the intestine of neonatal calves and cause fatal diarrhea. All of these strains have pili which are hair-like surface structures that bind to the intestine and anchor the bacteria against peristaltic flow. If calves are treated with monoclonal antibodies directed against pili immediately before infection, the severity of the resulting diarrhea is markedly reduced, presumably because the antibodies prevent anchoring of the bacteria and thereby reduce colonization.

TRANSGENIC ANIMALS

The term "Transgenic Animals" has been defined as those animals that have stably integrated into their germ line foreign DNA which had been introduced experimentally (Palmiter and Brinster, 1985). By this definition, the first transgenic animals were created in 1973 when the virus SV40 was introduced into the germ line of mice by injecting SV40 DNA into the blastocoel of early embryos (Jaenisch and Mintz, 1974). However, the experiment which brought transgenic animals to the forefront of scientific imagination and public interest was the experiment by Palmiter *et al.* (1982) in which mice were made transgenic with the rat growth hormone gene. The transgenic mice not only bore the rat growth hormone gene in their germ line but at least in some cases expressed it at high levels and grew to a truly impressive size. The extensions of this technology to animal production are obvious: genes could be introduced into the germ line of breeding stock which would confer virtually any desirable trait on the progeny including disease resistance, rapid growth, fecundity, and efficient utilization of feed. Because of the germ line inheritance of the traits, new traits could be added at the rate of a few per generation until one obtained an animal totally customized for its intended use.

At this point, it is instructive to compare the traditional approach to animal breeding with the transgenic approach. By the traditional approach, breeders have used the gene pool of the species, including rare mutations, to create animals bearing the characteristics they desired. By the transgenic approach, breeders will be able to use the collective gene pool of all living organisms as well as beneficial mutations which they have created to achieve the same end. Those who oppose transgenic animal breeding because it affronts the genetic integrity of the species should note the qualitative similarity between transgenic and traditional breeding and they should at the same time remember that contemporary stock animals are the products of gene pools which have been intensively manipulated since the dawn of domestication.

Two different approaches have been employed to produce transgenic animals. The Palmiter/Brinster approach involves direct intervention shortly after the egg is penetrated by sperm. At this time, the sperm's head rounds up to form the male pronucleus which contains the genetic contribution of the male to the offspring. At approximately the same time, meiosis is completed in the egg and the female pronucleus forms. The two pronuclei then fuse to form a diploid nucleus which is the first nucleus of the offspring. By the original Palmiter/Brinster approach, the recombinant DNA was microinjected into the male pronucleus immediately before fusion. Brinster *et al.* (1985) have subsequently shown that injection into the female pronucleus is nearly as efficient but injection into the cytoplasm surrounding the two pronuclei or injection into one of the two diploid nuclei immediately after the first division is considerably less efficient. That the DNA must be injected into the pronuclei gives rise to the only apparent limitation of the Palmiter/Brinster approach to creating transgenic animals: the pronuclei must be visualized to be microinjected. The pronuclei of sheep ova could be seen only by special microscopy and the pronuclei of swine ova were visible only after the egg's cytoplasm had been stratified by centrifugation (Hammer *et al.*, 1985b). These authors found that integration of a human growth hormone gene was about 10 percent efficient in pigs but only about one percent efficient in sheep.

The second approach to producing transgenic animals employs retroviruses, which can be integrated into the genome of an infected cell and inherited by progeny of that cell as a stable genetic trait. This property can be exploited by exchanging a portion of the viral genome for the gene of interest as was done by Jähner *et al.* (1985). These researchers infected mouse embryos at the four to eight cell stage with a construct of the retrovirus Moloney murine leukemia virus bearing the *E. coli* gene for the enzyme Xanthine (guanine) phorphoriboryl transferase. Of seven mice which were derived from infected embryos, one male was transgenic for the bacterial gene and his progeny were crossed to produce a strain of mice homozygous for this locus.

In a similar experiment, Souza *et al.* (1984) replaced the Src gene of the avian retrovirus, Rous Sarcomca virus, with the gene for chicken growth hormone and then infected nine-day-old chick embryos with the recombinant virus. After hatching, 50 percent of the chicks had circulating GH levels three- to ten-fold higher than normal but did not grow more rapidly.

Integration into the germ line was not tested and probably did not occur given the advanced age of the embryos when infected.

The retroviral approach to producing transgenic animals is clearly effective and in many ways more convenient than microinjection. However, any large-scale application of the retroviral approach to producing transgenic stock animals would result in the dissemination of genes from viruses which as a group are associated with such diseases as AIDS and cancer and will certainly be deferred until legitimate safety concerns have been addressed.

A potential technical problem with transgenic animals arises from the possibility that expression of a particular gene might be very desirable during one phase of an animal's life but very damaging during another. As an example, female mice transgenic for the growth hormone gene are usually infertile (Hammer *et al.*, 1985a). This general problem doubtlessly will be solved by including regulatory elements in the gene which put the gene under the control of the animal's own regulatory systems or which enable the gene to be turned on by simply adding an inducer to the animal's diet. An example of the latter is the growth hormone gene construct used by Palmiter and Brinster to produce transgenic mice. In this construct the growth hormone gene was placed under the control of the regulatory element for the metallothionine gene. Because the metallothionine gene is normally induced by zinc, growth hormone expression in the transgenic mice could be regulated by simply adjusting the level of zinc in their diet.

SUMMARY

THE PRESENT

Biotechnology has the potential to supply the animal industry with improved and novel diagnostics, vaccines, production improvers, therapeutics and even stock animals. In the case of diagnostics and vaccines, this potential has been partially realized. In the case of the growth hormones as production improvers, realization awaits only the development of appropriate supporting technology by formulation scientists. And in the cases of transgenic animals and therapeutics like the interferons and interleukins, more research is needed to define problems and potential.

THE FUTURE

It is indeed a tribute to recombinant DNA technology that it has, in slightly more than a single decade, seriously depleted science's storehouse of interesting genes to manipulate. As an example, the interferons took centuries to be discovered, decades to be characterized, and only a few years to be cloned. Even though the pace of science has accelerated markedly since the discovery of interferons fifty years ago, it still takes far longer to discover an important effect and to characterize the proteins and genes responsible than it does to clone those genes. Thus, contemporary biotechnology has given us more power over life's processes but it has not removed the major obstacle to exercising that power which remains, as always, our incomplete understanding of those processes. Increased understanding will come only from research of the most basic type and the majority of this research must be supported by public monies and done at universities and other nonprofit institutions. It cannot and will not be done by the corporate sector simply because there is in this world no stockholder, financial analyst, or C.E.O. who has the patience to support decades of expensive research whose potential cannot even be defined. Thus, the future progress of biotechnology will depend in a very direct way on the extent to which public funds support basic research and the traditional centers of basic research like universities and nonprofit institutes.

The related question concerns the type of training future scientists must have so that they will be able to perform this basic research and exploit its results. There have been serious comments that conventional biological disciplines will be swallowed by the combined discipline of biochemistry/molecular biology and equally serious comments that biochemistry/molecular biology will be swallowed by the conventional disciplines. In either case, the point is made that the biological sciences are converging at a rate unthinkable even five years ago, and the future scientist must have interdisciplinary training which will enable him to conceptualize a research problem at any intellectual level from intact animal to nitrogenous base pair. It is a clear and difficult challenge to the university system to produce scientists who have this breadth of training.

ACKNOWLEDGMENT

The author gratefully acknowledges the helpful comments of Dr. Dale Bauman (Cornell University), Drs. J. Egerton, L. Gordon, H. Hafs, G. Koo and J. Schmidt (MSDRL), and Mr. G. Barringer (MSD AGVET). The inclusive literature searches of Allyson Houck and Shaun Holland of MSDRL Information Services and the capable typing of June Zabel of MSDRL are also gratefully acknowledged.

REFERENCES

Ainsworth, A.J., T.L. Lester and G. Capley. (1985). Monoclonal antibodies to *Leptospira interrogans servovar pomona. Canad. J. Comp. Med.* 49: 202-204.

Asimov, G.J. and N.K. Krouze. (1937). The lactogenic preparations from the anterior pituitary and the increase of milk yield of cows. *J. Dairy Sci.* 20: 289-298.

Bachrach, H.L. (1985). New approaches to vaccines. *Adv. Vet. Sci. Comp. Med.* 30: 1-38.

Baile, C.A., M.A. Della-Ferra and C.L. McLaughlin. (1983). Performance and carcass quality of swine injected daily with bacterially-synthesized human growth hormone. *Growth* 47: 225-236.

Bauman, D.E., P.J. Eppard, M.J. DeGecter and G.M. Lanza. (1985). Responses of high-producing dairy cows to long-term treatment with pituitary somatotrapin and recombinant somatotrapin. *J. Dairy Sci.* 68: 1352-1362.

Boer, G.F. de, A.D. M.E. Osterhaus. (1985a). Application of monoclonal antibodies in the detection of ALV-gs antigen in the Dutch avian leucosis eradication scheme. *Tijds. Diergen.* 110: 842-844.

Boer, G.F. de, A.D. M.E. Osterhaus. (1985b). Application of monoclonal antibodies in the avian leukosis virus gs-antigen elisa. *Avian Path.* 14: 39-55.

Botstein, D., R.L. White, M. Skolnick and R.W. Davis. (1980). Construction of a genetic linkage map in man using restriction fragment length polymorphisms. *Amer. J. Human Genet.* 32: 314-331.

Brinster, R.L., H.Y. Chen, M.E. Trumbauer, M.K. Yagle and R.D. Palmiter. (1985). Factors affecting the efficiency of introducing foreign DNA into mice by microinjecting eggs. *Proc. Nat. Acad. Sci.* USA 82: 4438-4442.

Brouty-Bóye, D. (1980). Inhibitory effects of interferons on cell multiplication. *Lymphokine Rep.* 1: 99-112.

Brumby, P.J. and J. Hancock. (1955). The galactopoietic role of growth hormone in dairy cattle. *New Zealand J. Sci. Tech.* 36A: 447-436.

Chung, A.C.Y. and S.N. Cohen. (1974). Genomic construction between bacterial species *in vitro*: Replication and expression of *Staphylo-*

coccus plasmid gene in *Escherichia coli*. *Proc. Nat. Acad. Sci.* USA 71: 1030-1038.

Chung, C.S., T.D. Etherton and J.P. Wiggins. (1985). Stimulation of swine growth by procine growth hormone. *J. Animal Sci.* 60: 118-130.

Crouch, C.F., T.J.G. Raybould and S.D. Acres. (1984). Monoclonal antibody capture enzyme-linked immunosorbent assay for detection of bovine enteric cornavirus. *J. Clin. Micro.* 19: 388-393.

DeMaeyer-Guignard, J. and E. DeMaeyer. (1985). Immunomodulation by interferons: recent developments. *Interferon* 6: 69-91.

Derynck, R. (1983). More about interferon cloning. *Interferon* 5: 181-203.

Epstein, C.J. and L.B. Epstein. (1983). Genetic control of the response to interferon in man and mouse. *Lymphokines* 8: 277-301.

Findlay, G.M. and F.O. MacCallum. (1937). An interference phenomenon in relation to yellow fever and other viruses. *J. Path. Bact.* 44: 405-424.

Friedman, R.M. and S.N. Vogel. (1983). Interferons with special emphasis on the immune system. *Adv. Immun.* 34: 97-140.

Gamble, H.R. (1984). Application of hybridoma technology to the development of a diagnostic test for swine trichinosis. In *Hybridoma Technology in Agriculture and Veterinary Research* pp. 274-281.

Gery, I. and J.L. Lepe-Zuniga. (1984). Interleukin 1: uniqueness of its production and spectrum of activities. *Lymphokines* 9: 109-125.

Gusella, J.F., N.S. Wexler, P.M. Coneally, S.L. Naylor, M.A. Anderson, R.E. Tanzi, P.C. Watkins, K. Ottina, M.R. Wallace, A.Y. Sakaguchi, A.B. Young, I. Shoulson, E. Bonilla and J.B. Martin. (1983). A Polymophic DNA marker genetically linked to Huntington's disease. *Nature* 306: 234-238.

Hammer, R.E., R.L. Brinster and R.D. Palmiter. (1985a). Use of gene transfer to increase animal growth. Cold Spring Harbor *Symp. Quant. Biol.* 50: 379-387.

Hammer, R.E., V.G. Pursel, C.E. Roxroad Jr., R.J. Wall, D.J. Bolt, K.M. Ebert, R.D. Palmiter and R.L. Brinster. (1985b). Production of transgenic rabbits, sheep and pigs by microinjection. *Nature* 315: 680-683.

Holley, D.L., S.D. Allen, B.B. Barnett. (1984). Enzyme-linked immunosorbent assay, using monoclonal antibody, to detect enterotoxic *Escherichia coli* K99 antigen in feces of dairy calves. *Amer. J. Vet. Res.* 45: 2613-2616.

Hoskins, M. (1935). A protective action of neurotropic against viscerotropic yellow fever virus in macacus rhesus. *Amer. J. Trop. Med. Hyg.* 15: 675-680.

Isaacs, A. and J. Lindenmann. (1957). Virus interference I: The interferon. *Proc. Royal Soc.* B147: 258-267.

Jaenisch, R. and B. Mintz. (1974). Simian virus 40 DNA sequences in DNA of healthy adult mice derived from preimplantation blastocysts injected with viral DNA. *Proc. Nat. Acad. Sci.* USA 71: 1250-1254.

Jähner, D.K. Haase, R. Mulligan and R. Jaenisch. (1985). Insertion of the bacterial gpt gene into the germ line of mice by retroviral infection. *Proc. Nat. Acad. Sci.* USA 82: 6927-6931.

Jochim, M.M. and S.C. Jones. (1984). Identification of B.T. and EHD viruses by immunofluorescence with monoclonal antibodies. *Proc. Ann. Meet. American Association of Veterinary Laboratory Diagnosticians* 26: 277-286.

Johnsson, I.D. and I.C. Hart. (1985). The effects of exogenous bovine growth hormone and bromocryptine on growth, body development, fleece weight and plasma concentrations of growth hormone, insulin and prolactin in female lambs. *Animal Production* 41: 207-217.

Klausner, A. (1984). IML introduces interferon product for cattle. *Bio/Technology* October 1984, 841.

Kohler, G. and C. Milstein. (1975). Continuous cultures of fused cells secreting antibody of predefined specificity. *Nature* 256: 495-497.

Lachman, L.B. and A.L. Maizel. (1980). Human immunoregulatory molecules: interleukin 1, interleukin 2 and B-cell growth factor. *Contemp. Topics Molec. Immun.* 9: 147-167.

Levy, H.B. and F.L. Riley. (1983). A comparison of immune modulating effects of interferon and interferon inducers. *Lymphokines* 8: 303-322.

Lomedico, P.T., V. Gubler, C.P. Hellman, M. Dukovitch, J.G. Giri, Y-CE. Pan and K. Collier. (1984). Cloning and expression of murine interleukin 1 cDNA in *E. coli*. *Nature* 312: 458-462.

Maniatis, T., E.F. Fritsch and J. Sambrook. (1982). *Molecular Cloning, A Laboratory Manual* Cold Spring Harbor Laboratory, Cold Spring Harbor, NY.

Mattingly, J.A. (1984). An enzyme immunoassay for the detection of all *Salmonella* using a combination of a myeloma protein and a hybridoma antibody. *J. Immun. Meth.* 73: 147-156.

Mills, K.W. and K.L. Tietze. (1984). Monoclonal antibody enzyme-linked immunosorbent assay for identification of K99-positive *Escherichia coli* isolates from calves. *J. Clin. Micro.* 19: 498-501.

Morris, J.A., C.J. Thorns and C. Boarer. (1985). Evaluation of a monoclonal antibody to the K99 fimbrial adhesin produced by *Escherichia coli* enterotoxigenic for calves, lambs and piglets. *Res. Veter. Sci.* 39: 75-79.

Mosley, B., C.J. March, A. Larsen, D.P. Cerretti, G. Braedt, V. Price, S. Gillis, C.S. Henney, S. Kronheim, K. Grabstein, P.J. Conlon, T.P. Hopp and D. Cosman. (1985). The cloning and expression of two distinct human interleukin-1 (IL-1) cDNAs. In: *The physiological, metabolic and immunologic actions of interleukin-1* Alan R. Liss, NY. pp. 521-532.

Muir, L.A., S. Wien, P.F. Duquette, E.L. Rickes and E.H. Cordes. (1983). Effects of exogenous growth hormone and diethylstilbestrol on growth and carcass composition of growing lambs. *J. Animal Sci.* 56: 1315-1323.

Palmiter, R.D. and R.L. Brinster. (1985). Transgenic mice. *Cell* 41: 343-345.

Palmiter, R.D., R.L. Brinster, R.E. Hammer, M.E. Trumbauer, M.G. Rosenfeld, N.C. Birnberg and R.M. Evans. (1982). Dramatic growth of mice that develop from eggs microinjected with metallothionein-growth hormone fusion genes. *Nature* 300: 611-615.

Quinn, R., A.M. Campbell and A.P. Phillips. (1984). A monoclonal antibody specific for the A antigen of *Brucella* spp. *J. Gen. Micro.* 130: 2285-2289.

Rodriquez, R.L. and R.C. Tait. (1983). *Recombinant DNA Techniques: An Introduction* published by: Addison-Wesley Publishing Co., Reading, MA.

Scott, G.M. (1983). The toxic effects of interferon in man. *Interferon* 5: 85-114.

Sherman, D.M., S.D. Acres, P.L. Sadowski, J.A. Springer, B. Bray, T.J.G. Raybould and C.C. Muscoplat. (1983). Protection of calves against fatal enteric colibacillosis by orally administered *Escherichia coli* K99-specific monoclonal antibody. *Infection and Immunity* 42: 653-658.

Silva, R.F. and L.F. Lee. (1984). Monoclonal antibody-mediated immunoprecipitation of proteins from cells infected with Marek's disease virus or turkey herpes virus. *Virology* 136: 307-320.

Souza, L.M., T.C. Boone, D. Murdock, K. Langley, J. Wypych, D. Fenton, S. Johnson, P.H. Lai, R. Everett, R-Y. Hsu and R. Bosselman. (1984). Application of recombinant DNA technologies to studies on chicken growth hormone. *J. Exp. Zool.* 232: 465-473.

Stevens, A.E., S.A. Headlam, D.G. Pritchard, C.J. Thorns and J.A. Morris. (1985). Monoclonal antibodies for diagnosis of infection with *Leptospira interrogans* servovor hardjo by immunofluorescence. *Veter. Rec.* 116: 593-594.

Sutherland, S. (1985). An enzyme linked immunosorbent assay for detection of *Brucella abortus* in cattle using monoclonal antibodies. *Aust. Veter. J.* 62: 264-268.

Teramoto, Y.A., M.M. Mildbrand, J. Carlson, J.K. Collins and S. Winston. (1984). Comparison of enzyme-linked immunosorbent assay, DNA hybridization, hemagglutination, and electron microscopy for detection of canine parvovirus infections. *J. Clin. Micro.* 20: 373-378.

Weil, G.J., M.S. Malane, K.G. Powers and L.S. Blair. (1984). Monoclonal antibody-based assay for parasite antigenemia in *Dirofilaria immitis*-infected dogs. *Fed. Proc.* 43: 1667.

Wood, D.D., M.J. Staruch, P.L. Durette, W.V. Melvin III and B.K. Graham. (1983). In: *Interleukins, Lymphokines and Cytokines* Academic Press, NY. pp. 691-696.

Young, F.G. (1947). Experimental stimulation (galactopoiesis) of lactation. *Brit. Med. Bull.* 5: 155-160.

PART III
ADVANCES IN BIOTECHNOLOGY: ENVIRONMENT, HEALTH AND ENERGY

Impacts of Biotechnology on Biomedical Sciences

Thomas W. O'Brien
Department of Biochemistry and Molecular Biology
University of Florida
Gainesville, Florida

Researchers working in the biomedical sciences are well aware of the tremendous positive impact that biotechnology has had in this area. As illustrated in Figure 1, much of the biotechnology related research leading to medical advances involves recombinant DNA approaches, the production and utilization of monoclonal antibodies, and most recently, protein engineering. A most fortunate aspect of this research is that its very products have a direct positive feedback on research capabilities, providing new tools to carry out this kind of research. The rapid expansion of recombinant DNA methodologies following the introduction of DNA restriction

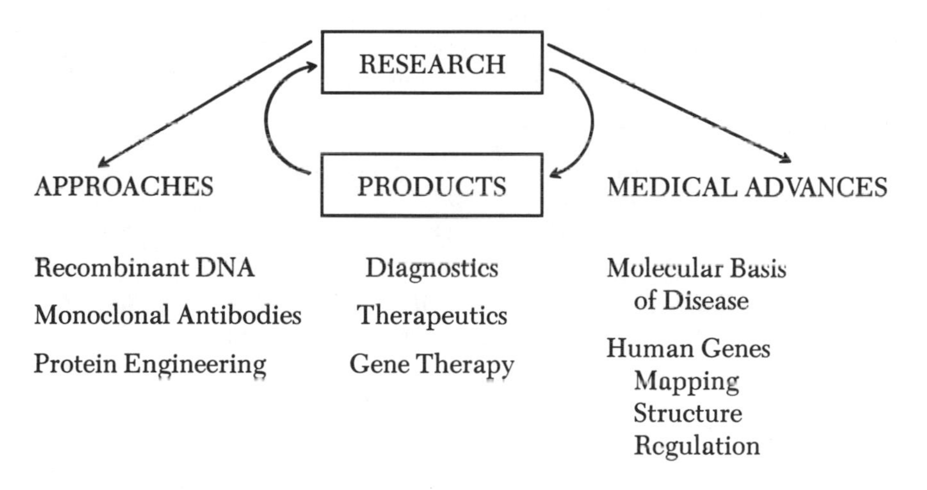

Figure 1. Impacts of biotechnology on the biomedical sciences. Medical advances and medical products of biotechnology are shown in relation to biotechnology research, emphasizing the posssible feedback of the intermediate products of biotechnology on research in this area.

enzymes is a good example of this process whereby early advances and products led to the development of new vectors and new research tools. This, in turn, speeds the discovery process.

Important medical advances are being made as a result of this rapid enhancement of biomedical research through biotechnology. With each advance comes increased understanding of the molecular basis of disease. Along these lines, encouraging progress is being made in the study of the human genome. Increased numbers of human genes are being identified and localized at high resolution on individual chromosomes. Finally, given knowledge of their localization and detailed structure, we will be in a better position to understand their regulation. Despite the optimism of the moment, it should be emphasized that we are only in the beginning stages of what promises to be a long and interesting endeavor.

The major products of biotechnology in the biomedical area fall into the broad categories of diagnostics and therapeutics. We shall consider these in turn, ending with the topic of gene therapy.

DIAGNOSTICS

The biotechnology-derived diagnostics are of two main kinds, monoclonal antibodies and DNA probes. Several companies, including Centocor, Cetus, Damon Biotech and Monoclonal Antibodies, among others, have major involvements in the production of monoclonal antibodies for diagnostic purposes. Over 150 different monoclonal antibodies are on the market now. Mainly used for diagnosis of various infectious diseases, they are also being used to detect various cancers and for pregnancy testing.

Because of their specificity and the constancy of their properties, monoclonal antibodies are ideally suited for several diagnostic applications. One of the real advantages of monoclonal antibodies, in addition to those discribed by Weppelman (this volume), is the fact that they are indeed derived from an individual clone. Thus, for example, hybridomas producing antibodies against cancerous cells can be cloned and screened to identify those producing antibodies against cell surface determinants that are found only on the cancerous cells, and not on normal cells. Such monoclonal antibodies, specific for different kinds of cancer cells, are now being produced not only for the detection and identification of different kinds of cancerous cells, but also for the localization of specific tumors. A recent example of the latter application involves the noninvasive imaging of ovarian carcinoma using monoclonal antibodies produced against the carcinoma-specific determinants by Centocor. Such antibodies can be tagged with an appropriate high energy emission isotope, such as ^{131}I, and administered to patients, using external gamma detectors to image the tumor.

DNA probes represent the other side of the coin in the diagnostics area. One of several commercially available DNA synthesizers can be used to make a DNA probe, a small stretch of DNA that is complementary to a known DNA or gene sequence from any source. Such oligonucleotides, designed to be specific hybridization probes for given segments of human, bacterial or viral DNA, can be labeled and used in a wide variety of applications. Various leukemias correlated with chromosomal translocations can be identified by this approach. In addition, oligonucleotide probes can expedite the prenatal diagnosis of certain genetic disorders for which specific DNA defects have been identified, using chorionic biopsy to obviate the lengthy process of cell culture following amniocentesis. Of course, bacterial diseases are also amenable to diagnosis using DNA probes. One of the recent successes in this area, announced in May, 1986, applies to the detection of bacteria causing Legionnaires disease. At least 22 different varieties of this bacterium exist, as identified immunologically, making diagnosis of all serotypes by antibodies impractical. However, oligonucleotide probes designed to be complementary to regions of the bacterial DNA which do not specify antigenic epitopes, and which are common to all strains, make it possible to detect all Legionnaires varieties in a single screening procedure.

THERAPEUTICS

As described by Abelson (this volume), some of the therapeutic products of biotechnology have already reached the market, albeit through a long and tortuous path. Recombinant insulin and human growth hormone are being marketed, and the FDA has finally approved (April, 1986) recombinant alpha interferon for marketing in the United States. These are listed, along with some others in the pipeline or on the drawing board, in Table 1.

RECOMBINANT THERAPEUTIC PRODUCTS

Insulin is the first recombinant human therapeutic product to reach the market. Developed by Genentech, recombinant insulin is now being produced under license by Lilly. A relatively simple molecule, recombinant human insulin is being produced in bacteria. Interestingly, insulin contains two amino acid chains. In the human, insulin is made in the form of a

Table 1. Examples of therapeutic products of biotechnology.

- Human insulin
- Human growth hormone
- Interferons
- Interleukins
- Growth factors
- Blood clotting factors
- Plasminogen activator
- Tumor necrosis factor
- Monoclonal antibodies and novel recombinant antibodies
- Synthetic peptide and recombinant vaccines
  - Hepatitis B
  - Malaria
  - Rabies
  - AIDS

single preproinsulin molecule, which is subsequently cleaved to give a single chain of proinsulin. The proinsulin molecule folds into a compact structure, stabilized by disulfide bridges. Enzymatic excision of a central loop in this folded chain gives rise to the mature insulin molecule, now consisting of the two smaller chains which are anchored together by disulfide bridges. Since bacteria lack the enzyme systems required to process preproinsulin, recombinant insulin was produced following a different strategy. Two separate kinds of DNA molecules were synthesized, each coding for one or the other insulin chain. After adding initiation codons (coding for methionine) to each, they were inserted into bacterial plasmid vectors, within the first part of the region coding for the protein beta-galactosidase. The mixture of fusion proteins produced in this manner can then be treated with cyanogen bromide, which cleaves them at the methionine residues which were engineered at the start of each chain of insulin. The released chains can then be combined to form insulin.

Since insulin is a relatively simple molecule, containing no methionine residues in the mature form, the above strategy is effective, using a bacterial system to produce the recombinant product. Bacterial systems are not well suited for making more complicated molecules which require secondary modification such as addition of carbohydrate groups and other reactions restricted to eukaryotes or even mammalian cells. In this case, alternative expression systems are available, either yeast cells or mammalian cells, if required.

The second example of a therapeutic product of recombinant DNA technology, human growth hormone, reached the market in October, 1985. This hormone, also developed by Genentech, is being produced in bacteria as well. The first recombinant product was identical to human growth hormone in all respects, including biological activity, except that it has a methionine as the first amino acid.

The third recombinant therapeutic product to be approved by the FDA for distribution and marketing in this country is alpha interferon. Developed by Genentech/Hoffman La Roche and Biogen, alpha interferon is also being produced in bacteria. The recombinant product appears to offer protection against cold viruses. It is anticipated that it will prove useful against a variety of cancers and viral diseases. In April, 1986, FDA approval was granted to Schering-Plough to market alpha interferon for treatment of hairy cell leukemia, multiple myeloma, non-Hodgkin's lymphoma and Kaposi's sarcoma.

Another interferon, gamma interferon, was also developed by Genentech. The recombinant gamma interferon can be produced in bacteria, and it is now in Phase III Clinical Trials by Biogen. Other companies involved in its production are Amgen and Schering-Plough. The recombinant product activates macrophages and it is expected to have important antitumor and antiviral properties.

Another lymphokine receiving much attention is recombinant interleukin II. This product was developed by Cetus, and several other companies, including Immunex, Biogen and Collaborative Research, which are also involved in its production. It is hoped that interleukin II will prove useful in treating immune deficiencies resulting from cancer, chemotherapy, childhood diseases and aging. A Japanese firm, the Takeda Chemical Company, recently announced the large scale production of interleukin II in bacteria. So much recombinant product is manufactured in the bacterium that large inclusion bodies of the protein form. It is indeed remarkable that scarce proteins such as interleukin II can now be produced in such quantities to allow physical studies of the molecule usually reserved only for abundant proteins.

A considerable market exists for blood proteins now produced by conventional blood fractionation procedures. One of the important blood clotting proteins, factor VIII, is present in such low concentrations that large volumes of blood must be processed to obtain this factor using conventional procedures. Absence or deficiency of this protein results in classical hemophilia A. Because of its scarcity and medical importance in the treatment of hemophilia, factor VIII was an obvious early candidate for application of recombinant technologies. Factor VIII is being produced by Genentech/Cutter, Chiron and Biogen. The production of recombinant factor VIII represents a major technological feat, since it is the largest protein gene cloned to date.

Another important therapeutic blood protein is tissue plasminogen activator (TPA). Developed by Genentech, it will also be produced by Biogen/Smith, Kline and Beckman, Damon Biotech and Chiron. TPA is glycosylated and it is being produced in mammalian cells, since no active products are obtained using bacterial systems. In tests of recombinant TPA, it appears to be very effective in dissolving blood clots, and it shows great promise for use in preventing strokes and ischemic damage following heart attacks.

Tumor necrosis factor (TNF), normally made by macrophages, is being produced as a recombinant molecule in bacteria by Genentech, Cetus, Biogen and Cellular Products. In clinical trials it appears to act synergistically with gamma interferon against various tumors. TNF treatment has some toxic side effects, however. It was recently discovered that TNF is the same factor as cachectin, a substance normally produced under stress and in response to endotoxin. Cachectin is associated with a general wasting effect in the body, which may account for the observed side effects of TNF treatment. Much remains to be learned about the specific roles and complex interactions of these important lymphokines and regulatory factors.

MONOCLONAL ANTIBODIES

Monoclonal antibodies are also being developed for use in therapy. Early efforts relied on monoclonal antibodies developed in rodent systems against specific human pathogens. While such antibodies are often therapeutically effective, they are of limited long term use because the patient eventually musters an immune response against the rodent proteins. For this reason, increasing use is being made of the recently developed human hybridoma systems for monoclonal antibody production. Alternatively, a novel recombinant approach is being taken by some investigators to combine the antigen-specific variable region of the successful rodent monoclonal antibody with the constant region of the human immunoglobulin molecule. This approach offers promise in reducing the side effects of the unique, disease specific monoclonal antibodies originally produced using rodent hybridomas.

In efforts to increase the effectiveness of monoclonal antibodies, they are also being combined with other agents to form specific immunotoxins. Because of their exquisite specificity, they are ideal vehicles for delivering attached toxins or radioactive molecules to specific cellular targets. Monoclonal antibodies specific for surface antigens in certain cancerous cells are being combined with various toxins and used for the selective killing of cancerous cells residing in the midst of normal cells. Agents such as ricin, the highly toxic protein synthesis inhibitor, are being used in this application, since a single molecule of such inhibitors is sufficient to inactivate all cytoplasmic ribosomes in the targeted cell. The toxin exerts little or no systemic side effects while attached to the antibody, but does become activated after the immunotoxin enters targeted cells by endocytosis. Variations on this theme include using immunotoxins as "proximity detonators" to release their toxins extracellularly in the vicinity of other tumor cells which may not yet be expressing the particular tumor antigen being targeted. This feature allows the destruction of neighboring tumor cells which are at an earlier stage in tumor progession.

Another novel approach to therapy, using monoclonal antibodies, is currently undergoing testing at the University of Florida Health Science Center. Developed by scientists at the Imperial Cancer Research Fund, this process uses monoclonal antibodies against tumor-specific cell surface proteins to purge bone marrow explants of tumor cells. In this process of immunomagnetic marrow purging, the monoclonal antibodies are used to coat magnetic microspheres which attach to the tumor cells, allowing their removal during passage through a magnetic trapping device. The purged bone marrow cells can be reintroduced to the patient after the patient's cancerous bone marrow has been destroyed by radiation and/or chemotherapy.

SYNTHETIC VACCINES

Much interest is presently directed towards the development of synthetic vaccines. Advances in protein structure determination, and in computer assisted molecular graphics, help biomedical scientists to identify potential antibody binding sites on the surface of proteins to be targeted by antibodies. One of the commercially available peptide synthesizers can then be used to make a peptide having the same amino acid sequence (and, hopefully, the same structure) as the targeted protein domain. When such peptides are coupled to carrier proteins, they very often elicit antibodies which recognize the targeted domain. An alternative approach to vaccine development utilizes recombinant DNA methods and the vaccinia virus as a vehicle for targeted proteins. This approach is being pursued by several biotechnology companies. Chiron, for example, has genetically engineered the vaccinia virus genome to express the hepatitis B antigen. Immunization with this recombinant vaccinia virus expressing the hepatitis B antigen is expected to confer immunity to hepatitis B. This recombinant DNA approach is also being used to engineer vaccinia virus constructs capable of expressing multiple antigens associated with different diseases. The ultimate goal of this approach is to provide protection against several diseases following immunization with multivalent immunogens.

Recombinant DNA technology also shows great promise in the quest for a malaria vaccine. Now that malaria is making a resurgence, and the malaria parasite is developing resistance to conventional therapeutic agents, an effective malaria vaccine is needed. Major attention is being directed towards the development of stage specific antibodies against the malaria parasite. It would also be desirable to be able to develop vaccines for any stage specific epitopes that are common to the different malaria species. Human trials of the first recombinant antimalarial vaccine began in March, 1986. The vaccine produced in *E. coli* is based on a cloned repeated sequence derived from the circumsporozoite gene of the malaria parasite. This vaccine was developed through collaborative efforts of scientists at Smith Kline and French and the Walter Reed Army Institute of Research.

HUMAN GENE THERAPY

Major reasons for considering human gene therapy are the severity and multitude of human genetic diseases. These therapies will not be rushed into service however, for there are certain risks inherent in this approach and many experimental details remain to be worked out. They will be undertaken only when the risk of harm is negligible. While biotechnological intervention is on the distant horizon for most of these diseases, some technically approachable diseases, which are severe and for which no successful therapy currently exists, are early candidates for recombinant gene therapy. For technical reasons, only those genetic diseases resulting from single gene defects will be targeted initially for gene therapy.

Human gene therapy is being approached through five main stages: (1) identification of the gene, (2) isolation of the gene, (3)analysis of the gene, to understand the nature of the problem, and (4) preclinical testing in non-human systems, before (5)gene insertion, modification or "gene surgery" in humans.

IDENTIFICATION OF GENETIC DISORDERS

Genetic disorders can be identified through linkage analysis and using specific DNA probes to characterize restriction fragment length polymorphisms (RFLP). Gross insertions or deletions of genetic information can be detected by Southern hybridization on samples of genomic DNA. Specific mutations can be detected using synthetic oligonucleotide probes spanning the mutation site, as well as restriction enzymes which cut within this site. It is technically possible to obtain precise information about the mutation site through direct analysis with synthetic oligonucleotides.

PROSPECTS FOR GENE THERAPY IN THE NEAR FUTURE

The Office of Technology Assessment Background Paper on Human Gene Therapy identifies five candidate genetic diseases for initial consideration. The severe immunodeficiency caused by the enzyme *adenosine deaminase (ADA) deficiency* is one of these. Approximately fifty cases of ADA deficiency have been reported worldwide. Another candidate genetic disease, *purine nucleoside phosphorylase deficiency*, has been identified in nine patients in six different families worldwide. Another, more common genetic disease, the *Lesch-Nyhan Syndrome*, affects one in ten thousand males, resulting in complete hypoxanthine-guanine phosphoribosyl transferase deficiency. It is estimated that two hundred new cases of this disease occur each year in the United States. *Citrullinemia* is another candidate disease, with over fifty reported cases of urea cycle defects caused by deficiencies of the enzyme arginosuccinate synthase. Finally, over one hundred cases of *ornithine carbamoyl transferase (OCT) deficiency* have been reported.

Other genetic diseases for which gene therapy may be possible in the foreseeable future include phenylketonuria (as an improvement on the current dietary treatment), familial hypercholesterolemia, and defects of the urea cycle other than citrullinemia and OCT deficiency, such as arginase deficiency. In addition, some of the mucopolysaccharidoses and other defined metabolic defects may be considered. These include some forms of Gaucher's Disease, metachromatic leukodystrophy (arylsulfatase B deficiency with little brain involvement), the Hunter Syndrome and severe grades of branched chain ketoaciduria.

Gene therapy may not be applicable for other conditions such as chromosomal disorders, like Down's Syndrome, which involve non-disjunction or translocation of chromosomal segments. Others in this category include environmental and multigenic disorders such as hypertension and diabetes, and complex traits such as physical strength or intelligence.

Promising vectors being developed for human gene therapy are likely to be modeled after retroviruses like Moloney murine leukemia virus (MLV).

The basic strategy to be followed in engineering this vector will be to replace the viral structural genes with the desired human gene, complete with its own promoter. Because this construct will lack viral sequences necessary for packaging the RNA into infectious particles, it must be replicated in the presence of a helper MLV, which is, itself, rendered packaging deficient by deletion of a specific segment of the helper virus genome. The infectious vectors obtained through this approach will be used to deliver the selected human gene to some of the patient's bone marrow cells *in vitro*. After culturing the marrow cells to allow insertion of the new gene into cellular DNA, the cells will be returned to the patient. Note that these efforts are directed at gene therapy in somatic cells. The genetic alterations are thus non-heritable. Nevertheless, there is reason for optimism in expectations for the reversal or amelioration of some of these devastating genetic diseases.

SUMMARY

From the above overview, it should be apparent that we are only in the beginning stages of a new age of biomedical research. Furthermore, it is evident that the promises of applied biotechnology will not be realized without considerable additional basic research that forms the foundation for biotechnology. It should also be appreciated that the positive feedback noted between the intermediate products of biotechnology and biotechnology research extends to the more "traditional" biomedical areas as well. The field of protein chemistry, for example, is undergoing a vigorous expansion as a result of the new technologies and the need to purify, analyze and characterize the plethora of new biological products finally being produced in palpable amounts. Obviously, producing large amounts of scarce biological products and regulatory molecules is not enough. We must turn to the biochemists, physiologists, pharmacologists, microbiologists, cell biologists and pathologists in order to understand their complex interplay and develop them for the benefit of all mankind.

REFERENCES FOR FURTHER READING

Abelson, P.H., ed. (1984). *Biotechnology and Biological Frontiers*. American Association for the Advancement of Science, Washington, D.C.

Kidd, V.J. and Woo, S.L.C. (1984). Molecular diagnosis of human genetic disease. *Modern Cell Biology* 3: 113-129.

U.S. Congress, Office of Technology Assessment (1984). *Human Gene Therapy: A Background Paper*, OTA-BP-BA-32. U.S. Government Printing Office, Washington, D.C.

U.S. Congress, Office of Technology Assessment (1984). *Commercial Biotechnology: An International Analysis*, OTA-BA-218. U.S. Government Printing Office, Washington, D.C.

U.S. Congress, Office of Technology Assessment (1986). *Technologies for Detecting Heritable Mutations in Human Beings*, OTA-H-298. U.S. Government Printing Office, Washington, D.C.

Walters, L.R. (1986). The ethics of human gene therapy. *Nature* 320: 225-227.

Watson, J.D., Tooze, J. and Kurtz, D.T. (1983). *Recombinant DNA: A Short Course*. Scientific American Books, New York, N.Y.

Yu, S.-Y., von Ruden, T., Kantoff, P.W., Garber, C., Seiberg, M., Ruther, U., Anderson, W.F., Wagner, E.F. and Gilboa, E. (1986). Self-inactivating retroviral vectors designed for transfer of whole genes into mammalian cells. *Proc. of the Nat. Acad. of Sci.* U.S.A. 83: 3194-3198.

Impacts of Biotechnology on Marine and Environmental Sciences

Rita R. Colwell
Department of Microbiology
University of Maryland
College Park, Maryland

Dramatic changes are taking place in oceanography and marine sciences, and the economic value of research underway in applying biotechnology to these fields should not be understated. Shellfish harvests represent income of millions of dollars each year for coastal states, including Florida, California, Texas, and Maryland. Worldwide, notably in Asian countries, the fisheries industries represent $300 to $400 million each year in production of seafood and marine products. The state of Maryland has a special interest in marine biotechnology because of its major natural resource, the Chesapeake Bay. Landed harvests of Chesapeake Bay fisheries can be estimated at about $13 million, and significantly more, if one takes into account all of the auxiliary activities on the Chesapeake Bay and their value to the state. The economic value, therefore, approaches the billion dollar a year range. Thus, fisheries, as an industry, is significant on an economic basis, even without considering fish as a source of protein for millions of people each year.

A few examples of the value of marine biotechnology in enhancing fisheries' production and the maritime industries will be discussed in this paper. Excellent examples of the potential of marine biotechnology applied to the shellfish industry include the work of Dr. Daniel Morse and his colleagues at Santa Barbara, and the oyster studies underway at the University of Maryland. In these cases, genomic libraries for the abalone and oyster, respectively, have been established.

In addition to genetic engineering of shellfish, a great deal of work is being done on fish. Although the latter is not covered in detail here, it must be pointed out that genetic engineering of fish represents a very active area of research. Japanese scientists have reported successful cloning of the growth hormone in salmon. Scientists working in California have also reported, independently, the cloning of the salmon growth hormone. At Johns Hopkins University and the University of Maryland, a team of scientists have cloned the growth hormone in trout. A group of scientists in Vancouver, Canada, demonstrated the use of growth hormone to obtain rapid growth and larger size of salmon. These advances represent only the beginning of a very exciting era in marine biology. Clearly, much has been happening in marine biotechnology.

Seaweeds represent a massive source of an economical food product for many countries of the world. Aquaculture of seaweed has been underway for several years, but new advances are being made in the application of molecular genetic methods to seaweed culture. Genetic engineering of seaweed is just beginning to get underway.

Marine genetic engineering also offers potential technological advances, such as development of new, innovative biosensors. Marine biosensors, in particular, offer interest because chemicals produced by marine organisms frequently serve as attractants, representing "model" sensors. Lipids produced by deep sea bacteria have also proven to be interesting and may, in the future, provide useful and novel biosensors.

Marine pharmacology is an area that is moving very rapidly, with molecular genetic advances being made. For example, genes controlling production of venom in the sea snake have been cloned. Extensive screening of marine organisms for compounds of pharmaceutical application is currently underway and represents an area of special interest.

The need, at the present time, if the potential of marine biotechnology is to be fulfilled, is for significantly more basic research on the molecular biology of marine organisms. Marine bacterial genetic systems need to be developed, to augment the work done to date and to provide vectors and cloning systems for marine biology.

Marine environmental engineering represents an aspect of marine biotechnology that is equally important and offers promise in dealing with the problems of environmental pollution. Biotechnology-based disposal systems for the marine environment are beginning to be developed. It makes little sense, for example, to discharge fresh-water wastes to the open ocean and expect biodegradation to proceed rapidly. In fact, the high salt, low temperature, and low nutrient conditions of the marine environment are inhibitory for fresh-water organisms, notably the bacteria. As the discharged material sinks to great depths in the ocean, hydrostatic pressure increases, further inhibiting biodegradation. Thus, for deep ocean dumping of wastes, biological systems should be developed that will allow biodegradation to proceed at a reasonably rapid rate after dumping at sea.

Biofouling is an activity of tremendous economic cost to marine industries, especially the commercial shipping industry, as well as the U.S. Navy. Biofouling of ships' hulls can increase impedance, i.e., drag, significantly, in a matter of days or months. In an effort to understand the fundamental nature, i.e., the initial steps of biofouling, several scientific teams are engaged in some very exciting work. An example of which, is that of Dr. M. Simon's group at the Agouron Institute, San Diego, California, who have cloned genes controlling production of lateral flagella in bacteria involved in initial stages of biofilm formation on surfaces. The processes involved in biofilm formation are being studied by a number of investigators and within the next decade, biotechnological control of biofouling can be expected.

Exciting opportunities are offered in the application of biotechnology to fish stock assessment. As one example, high seas fisheries management can effectively employ mitochondrial nucleic acid restriction patterns to identify and monitor fish populations. This molecular biology "fingerprint" provides a biological tag that unequivocally can identify the source of the fish, i.e., their geographical origin. Tuna and salmon populations can be readily identified with such methods, an important advance that can aid in establishing fishing rights and settling international disputes over fisheries issues.

Nucleic acid sequencing, specifically 16S and 5S RNA sequencing, has been effectively applied to procaryotes, as well as eucaryotes, including fish species, as a means of species classification and identification and in tracking evolution at the molecular level. Many new species have been discovered in the deep ocean in recent years. These methods will allow more precise identification and a means of understanding their phylogenetic ancestry.

In view of the potential of marine biotechnology, it is disheartening to learn the relatively small investment being made by the federal government in basic research in this area of science. Approximately $8 million per year is spent by the National Sea Grant College Program on marine biotechnology. The National Science Foundation has recently initiated a program in marine biotechnology. Thus, for the next few years, the amount of funding for marine biotechnology research will be *ca* $10 million or less.

The opportunities for successful research projects in marine biotechnology are really very good. Specific examples of research findings and work underway include the following. The spawning of the abalone is now reasonably well understood, based on the work of Morse and his colleagues. Addition of hydrogen peroxide to adult abalone ready to spawn will induce spawning, and allow synchronous development of the larvae. Achieving essentially synchronous culture of the abalone provides a major step forward in being able to develop an aquaculture system for the abalone. The abalone is found in association with crustose red algae. The algae release gamma amino butyric acid (GABA) mimetic, which acts as an attractant for the larvae of the abalone, inducing the larvae to settle and undergo metamorphosis. The University of California, Santa Barbara research group is in the process of cloning and amplifying the growth hormone, in order to rear the abalone in closed culture to larger size and at a rapid growth rate. The possibility, then, is that more abalone can be brought to market in a shorter time.

At the University of Maryland, the Chesapeake Bay oyster, a delicacy and an economically important shellfish, has been under study for a number of years because of declining harvests. We have found that the oyster, once fertilized and in the early larval stage, when "competent," will set. During search for an appropriate surface onto which to settle, the oyster extends its foot and, by means of a sensing organ, detects a chemical produced by bacteria. A bacterium present on such surfaces has been identified and shown to produce the surface film onto which the oyster larva sets, and the larva metamorphosizes to the "spat" state. The oyster then grows into an adult.

Larva require the microbial film for settlement and have developed a probable symbiotic relationship with the bacterium that has been found in association with the oyster. A chemical, produced by the bacterium strain LST, has been identified, and appears to be L-DOPA or L-DOPA mimetic, a neurotransmitter precurser, which, when released by the bacterium, is "sensed" by the larva. The bacterium also produces an abundant, highly viscous polysaccharide which attaches the bacterium to the surface. The polysaccharide appears also to play a role in larva settlement, possibly triggering cementation. In the metamorphosis of the oyster, epinephrine and norepinephrine are required. We now know that the bacterium LST produces an L-DOPA-like compound, originally discovered because the bacteria produce a pigment, subsequently determined to be melanin, for which it is a L-DOPA precurser. We have been able to confirm that a much greater oyster set is obtained if the oyster larvae are presented with surfaces coated with the bacterial film. Thus, we are at the point of being able to control production of the oyster in the hatchery, which up until now, was mostly a matter of chance. Field tests have been done, in which oyster shells coated with L-DOPA, or the bacterial film, were placed in aquaculture systems, both in the Chesapeake Bay, Maryland, and in Washington on the West Coast. Spat set was significantly increased when either L-DOPA or the bacterial film was present. A patent has been filed for the process, since it appears to have applicability in oyster culture and, possibly, in reversing the decline in oyster production in the Chesapeake Bay.

The bacterial polysaccharide has been found to be highly stable in seawater and this bacterial "glue" is being developed further, in collaboration with an industrial firm, as a potential "marine glue."

Experiments are now being done on cloning the tyrosinase gene(s), in order to develop a biotechnological source of L-DOPA for use in oyster hatcheries. A genomic library for the oyster has been established, with the intent of cloning the oyster growth hormone, for the same reasons that the salmon and trout growth hormone genes have been cloned, namely to speed growth and obtain larger size of the animals.

Seaweeds offer an interesting challenge because seaweed culture remains relatively unsophisticated, i.e., seaweeds are harvested from the sea or in culture using mechanical rigs. The seaweeds, once harvested, are washed, extracted chemically, and purified. Many commercially important chemicals are obtained from seaweeds, especially gelling agents, i.e., marine colloids, including agar and carrageenan. The marine colloids represent a multi-million dollar a year product for the food industry. Thus, a reliable source of agar and carrageenan, totally controlled in production, would be of significant commercial value.

Protoplasts of seaweeds have been produced and protoplast fusion has been accomplished. By this technique, seaweed hybrids have been formed with the intent to achieve fast-growing, abundant colloid-producing strains. In the long run, the molecular genetics of seaweeds must be established, but it has only just begun, in the United States, Japan, and China. The prospects are excellent for a biotechnologically developed, commercial source of marine colloids in the future.

Marine pharmacology is another fascinating application of marine biotechnology. However, a molecular genetic understanding of marine biological systems, including the chemical mechanisms of territorial circumscription and biochemical ecology is needed. For example, the structure and function of marine toxins are now being detailed.

Systematic studies of marine organisms have revealed potential anti-cancer agents as early as the mid-1960s by Petit and others. Screening for anti-cancer compounds produced by marine organisms is underway at a number of universities, including the University of Oklahoma, University of Miami, University of Illinois, and more recently, the University of Maryland. Several tumor inhibitory compounds have been described in the literature over the years, with the most active anti-tumor compounds from marine organisms reported since 1981. Rinehart and his co-workers have discovered some interesting compounds, some of which are being considered for clinical trial by the National Cancer Institute.

Tumor inhibitory cyclic peptides have been isolated by several investigators during the past two decades. A list of such compounds can be obtained from the references cited below.

Japanese workers have focused on marine bacteria as a source of new antibiotics. Some bacteria associated with corals and marine sponges appear also to produce polysaccharides active as anti-tumor agents. Antimalaria and anti-leukemic agents have been isolated from marine bacteria. Potentially pharmacologically useful agents include tetrodotoxin, recently discovered by Japanese workers to be produced by bacteria found on the skin of certain fishes. Because it appears that the tetrodotoxin is produced by bacteria, there is an excellent opportunity to clone the genes involved in tetrodotoxin production.

Tetrodotoxin is similar to local anesthetics in action, viz., procaine and cocaine, in that it selectively blocks cell membrane transport. It has been claimed that tetrodotoxin is *ca* 300,000 times more potent than cocaine. Thus, as an analgesic, tetrodotoxin offers interesting prospects.

Palytoxin appears to have pharmacological value, as do a number of other toxins, such as lophotoxin and toxins from *Aplysia*, the sea hare, and other marine animals.

The tunicate, a primitive chordate of the family Didemnidaceae have been shown to produce compounds, the didemnins, active against the virus, *Herpes simplex* HSV1 and the P388 leukemia virus. When tested in mice, survival time is significantly extended when very small doses are applied. Thus, the didemnins may be a valuable anti-leukemic agent.

Totally unexplored is the potential of deep sea organisms, including bacteria, as a source of novel proteins and other macromolecules. Growth of bacteria at pressures of 165 psi or greater, and temperatures of up to 250°C has been reported. Specifically, the bacteria found at the hydrothermal vents of the deep ocean, such as the East Pacific Rise off the West Coast of the United States and at the submarine volcanic vents off the Galapagos Islands are of greatest interest. Bacteria have been isolated from the black smokers, i.e., submarine volcanic vents which emit very hot water. Such bacteria may possess novel protein structures, i.e., enzymes and other cell components. Thus, exploration and study of these bacteria, although only just beginning, offer extraordinary opportunities for commercially valuable products.

In studies done at the University of Maryland, plasmids have been found in deep sea bacteria, including strains isolated from the deep ocean trenches and hydrothermal vent regions. Plasmids appear to occur frequently in the deep sea bacteria, allowing some very interesting work to be done in the future. A bacterium that grows best under elevated hydrastatic pressure has been found to harbor a plasmid and plasmid transfer under pressure appears to take place quite readily. Thus, high pressure, high temperature marine systems, as well as high pressure and low temperature ecosystems of the world oceans, offer some very interesting biological material for study with excellent potential for marine biotechnology.

In summary, only a few of the potential applications of marine biotechnology have been discussed here and only very generally. Obviously, there are problems to be resolved. One has already been mentioned, namely the inadequate funding of marine biotechnology in the United States. *Ca* $8 to $10 million is spent, mainly by the National Sea Grant College Program, while the Japanese government is reported to be investing $16 to $30 million per year in marine biotechnology. Related to the lack of funds for research, is the need for trained marine biotechnologists, individuals knowledgeable about high temperature, high pressure, and other extreme environmental conditions, including anaerobic, high temperature, and high pressure systems which are technically difficult to work with and expensive to maintain. Given available funding and well-trained scientists, there is no doubt that some fascinating and economically useful work will be done.

In conclusion, marine and environmental biotechnology are truly exciting areas in which to work at this time. In the United States, advantage must be taken of the many opportunities offered by application of the tools of genetic engineering to marine and environmental biotechnology, areas of cutting edge science. The window of opportunity is open but without investment of resources, leadership will be lost.

REFERENCES FOR FURTHER READING

Bakus, C.J., N.M. Targett, B. Schulte. (1986). Chemical ecology of marine organisms: an overview. *Journal of Chemical Ecology*, 12(5).

Belas, R., M. Simon, M. Silverman. (1986). Regulation of lateral flagella gene transcription in *Vibrio parahaemolyticus*. *Journal of Bacteriology*, p. 210-218.

Baloun, A.J. and D.E. Morse. (1984). Ionic control of settlement and metamorphosis in larval *Haliotis Rufescens* (Gastropoda). *The Biological Bulletin*, 167(1).

Colwell, R.R. and J.R. Greer. (1986). Biotechnology and the sea. *Ocean Development and International Law*, 17(1/2/3).

Colwell, R.R., E.R. Pariser, A.J. Sinskey. (1984). Biotechnology in the marine sciences. *Proceedings of the First Annual MIT Sea Grant Lecture and Seminar.*

DePauw, N., J. Morales, G. Persoone. (1984). Mass culture of microalgae in aquaculture systems: progress and constraints. *Hydrobiologia*, 116/117.

Hodgson, L.M. (1984). Antimicrobial and antineoplastic activity in some South Florida seaweeds. *Botanica Marina* 27:387-390.

Imhoff, J.F., H.G. Truper. (1976). Marine sponges as habitats of anaerobic phototrophic bacteria. *Microbial Ecology* 3:1-9.

Kellogg, S.T., D.K. Chatterjee, A.M. Chakrabarty. (1981). Plasmid-assisted molecular breeding: new technique for enhanced biodegradation of persistent toxic chemicals. *Science*, 214.

Look, S.A., W. Fenical, R.S. Jacobs, J. Clardy. (1986). The pseudopterosins: anti-inflammatory and analgesic natural products from the sea whip *Pseudopterogoria elisabethae*. *Proc. Natl. Acad. Sci. U.S.A.* 83:6238-6240.

Marderosian, A.D. (1969). Marine pharmaceuticals. *Journal of Pharmaceutical Science*, 58:1-33.

Morse, D.E. (1984). *Biochemical control of larval recruitment and marine fouling*. Naval Institute Press, Annapolis., pp. 134-140.

Morse, D.E. (1984). Biochemical and genetic engineering for improved production of abalones and other valuable molluscs. *Aquaculture*, 39:263-282.

Morse, N.C., D.E. Morse. (1984). GABA-mimetic molecules from *Porphyra (Rhodophyta)* induce metamorphosis of *Haliotis (Gastropoda)* larvae. *Hydrobiologia* 116/117.

Morse, N.C., D.E. Morse. (1984). Recruitment and metamorphosis of *Haliotis* larvae induced by molecules uniquely available at the surfaces of crustose red algae. *J. Exp. Mar. Biol. Ecol.*, 75:191-215.

Sekine, S., T. Mizukami, T. Nishi, Y. Kuwana, A. Saito, M. Sato, S. Itoh, H. Kawauchi. (1985). Cloning and expression of cDNA for salmon growth hormone in *Escherichia coli*. *Proc. Natl. Acad. Sci.*, 82:4306-4310.

Tamiya, T., A. Lamouroux, J.R. Julien, B. Grima, J. Mallet, P. Fromageot, A. Menez. (1985). Cloning and sequence analysis of the cDNA encoding a snake neurotoxin precursor. *Biochimie*, 67:185-189.

Weiner, R.M., R.R. Colwell, R.N. Jarman, D.C. Stein, C.C. Somerville, D.B. Bonar. (1985). Applications of biotechnology to the production, recovery anduse of marine polysaccharides. *Bio/Technology*, 3:899-902.

Youngken, Jr., H.W., Y. Shimizu. (1975). Marine drugs: chemical and pharmacological aspects. *Chemical Oceanography*, 4:269.

Impacts of Bioengineering on Biotechnology

Alan Sherman Michaels
Department of Chemical Engineering
North Carolina State University
Raleigh, North Carolina

THE SCOPE OF BIOENGINEERING

While I am sure that it needs no redefinition here, I wish to describe "biotechnology" in my own terms as the utilization of living organisms, subcellular organelles, or biocatalysts (enzymes) for the synthesis of substances (bioproducts) which can benefit society. Within this context, I shall then define "bioengineering," for the purposes of my analysis, as the development and application of devices, processes, and systems which permit the safe, economic, and efficient manufacture and use of bioproducts.

Bioengineering has been a recognized sub-specialty of the engineering profession for many decades, largely housed within the core-disciplines of chemical, mechanical and electrical engineering, and materials science. Two principal branches of bioengineering are recognized today. They are: (1) **Biochemical/Bioprocess Engineering**. Addressing industrial-scale biosynthesis (fermentation, cell-culture, enzymatic transformation); separation/purification/isolation of bioproducts; energy- and waste-management in bioprocessing; and systems design and process control. (2) **Biomedical Engineering**. Addressing the techniques of fabrication and use of devices and materials, and more recently the methods of delivery and administration of bioproducts, for the treatment or prevention of human and animal diseases or disability.

The focus of my attention in this analysis will be on biochemical and bioprocess engineering as it relates to biotechnology. Limitations of space and time militate against my dealing in depth with biomedical engineering and its impact on biotechnology, despite its equivalent importance. Suffice it to say that the development of delivery systems for the safe and efficient administration of health care bioproducts to humans and animals is the key to their successful commercialization, and that such development falls squarely in the domain of the biomedical engineer.

Without question, the successful, large-scale application of modern biotechnology to the fulfillment of well-defined societal needs (whether it be the production and use of human growth hormone for the treatment of dwarfism, or the production and distribution of fermentation alcohol for use as a liquid fuel) hangs upon the availability of proper engineering skill and talent to design and develop the necessary processes and systems for the safe and economic manufacture of the desired products. Bioengineering is the critical bridge between modern biological science and the commercialization of bioproducts, in the same way that chemical engineering is the bridge between chemistry and industrial chemical production, or that electrical engineering is the bridge between solid state physics and the commercialization of electronic devices. It is, therefore, quite appropriate to deal with the impact of bioengineering on biotechnology in a symposium addressing the future of biotechnology, since without this engineering linkage, biotechnology would be a technology without a future.

Historically, the role of biochemical and bioprocess engineering in support of the life and medical sciences has been confined to the development, design, and construction of manufacturing plants for production of fermentation products (e.g., antibiotics, alcohol), of synthetic pharmaceuticals and nutritional chemicals, and of biological substances extracted from plant and animal tissues and body fluids (steroids, alkaloids, immunoproteins). The revolution in life science which has transpired over the past 25 years, and has made "biotechnology" a household word, has fundamentally altered the concept and scope of "engineering" as it applies to the manufacture and use of bioproducts. Traditional chemical and biochemical engineering are, as will be shown, inadequate to cope with the special problems and requirements for industrialization of modern biology, a realization that has been acknowledged by the engineering profession, as well as the financial and industrial community, only within the past few years. Only quite recently has the academic engineering community begun to respond to the needs of biotechnology by reappraising its teaching and research programs in biochemical/bioprocess engineering, and restructuring and vitalizing these programs to meet current and future industrial requirements.

In no small measure, the delay in recognizing and acknowledging the need for innovative engineering input in achieving the societal goals of modern biology had its origins in the reluctance of the life science research community to recognize the importance of the engineering disciplines to successful reduction-to-industrial practice of the dramatic discoveries of genetics and molecular biology. This reluctance was not a mark of malice on the part of the biological scientific community, but rather a measure of the naive unawareness of the basic life scientist that "getting the product from the cell into the bottle" involved considerably different expertise and technical know-how from those employed in the analytical and preparative techniques of the life science research laboratory. Regrettably, the captivation of the financial and business communities by the exciting discoveries in biology during the seventies and early eighties resulted in the birth of a biotechnology industry long on sophisticated life science, but very short on bioengineering competence. The meteoric rise of interest in this fledgling industry, and the dismal period of investor disenchantment which soon followed, were largely attributable to this deficiency in engineering perspective and capability. Only in the past few years has reason begun to prevail in assessing the commercial future of biotechnology, as the role of bioengineering is more fully appreciated.

THE ROLE OF ENGINEERING IN BIOTECHNOLOGY

The successful commercialization of modern biological scientific discoveries (like that of any industrial product) requires meeting the following objectives:

(1) The ability to manufacture the product on a scale commensurate with the anticipated market.

(2) The ability to produce the product at a level of purity, quality, and consistency satisfying competitive marketing and regulatory (product safety/efficacy) requirements.

(3) The ability to produce the product at a cost-of-manufacture and distribution which will assure acceptable market penetration and profit.

(4) The ability to produce the product in a manner involving acceptable hazards to operating personnel and impact on the environment.

These are classic engineering objectives, generally appreciated, for example, by a competent chemical engineer faced with the responsibility for commercializing a chemical product. Such an engineer is equipped with the training and professional tools required to design, develop, select, and integrate the process steps and hardware needed to build and operate a manufacturing facility for that product. This is the essence of process engineering, as it is widely and competently practiced around the globe today.

Inasmuch as biotechnology products are chemical substances, one might ask why the classically-trained chemical engineer is not qualified to perform all the engineering tasks needed to reduce these products to industrial practice. The answer is that the methods of synthesis of most genetically-engineered bioproducts, their physical and chemical properties, and the processes required to isolate and purify them, are so radically different from conventional chemicals and chemical process operations that the traditional chemical engineer is at best poorly equipped to address these problems. The same situation confronts the classically-trained biochemical engineer, whose facility in dealing with conventional fermentation processes and fermentation product recovery methods is of limited utility in solving the formidable manufacturing problems of many of the newer and most promising bioproducts. This is partly because existing methods of large-scale fermentation and cell culture are poorly suited to the industrial-scale growth and maintenance of recombinant organisms and mammalian or plant cells, which are fragile, fastidious in their environmental requirements, lacking in vigor, and slow producers. Moreover, most bioproducts directed to human- and animal-health care applications, such as peptide hormones, immunoproteins, cytokinetic oligopeptides and enzymes, are extremely complex and unstable molecules which are present in very low concentration in solution, admixed with large numbers of closely related compounds (which are usually unwanted or toxic impurities), from which they must be isolated in exceedingly high purity. Few existing downstream bioprocessing techniques can meet these imposing requirements.

In addition, many of these health-care biologicals are of extraordinary potency and biological activity, and many of the byproducts of their synthesis constitute a serious health hazard if allowed to escape into the environment in even minuscule amounts. Hence, the design and operation of plants for the manufacture of such products pose problems of product- and byproduct-containment, and of waste decontamination, unmatched in complexity in industry today.

Lastly, when one contemplates the ultimate manufacture of tonnage-quantities of industrial chemicals and gaseous or liquid fuels by biological routes, one faces the need to design and develop bioreactors capable of continuous cell-culture on a scale of unprecedented magnitude, separation/purification processes of immense capacity and unheard-of energy efficiency, and waste treatment facilities of massive size and complexity.

Thus the role of engineering in biotechnology, and its criticality in assuring the achievement of the societal goals which modern biological science promises, is clear. The vital questions to be answered are (1) whether engineers, life scientists, and business managers are prepared to collaborate constructively to attain these goals, and (2) how such collaboration need be structured and supported to assure that these goals will be reached rapidly and with minimum cost to society.

THE STATE OF THE ART IN BIOCHEMICAL/ BIOPROCESS ENGINEERING FOR BIOTECHNOLOGY

During the early years of the biotechnology revolution, the life scientists largely ignored the importance of engineering in their planning, and the engineers remained blissfully unaware of the problems the life scientists were beginning to encounter in preparing for the manufacture of bioproducts. This soon led to a situation where many of the start-up biotechnology ventures had perfected the genetic and biological laboratory protocols for producing exciting diagnostic and health-care products whose market potential was enchanting, but the means for their manufacture, even for preliminary clinical evaluation, was yet to be developed. This dilemma precipitated two courses of action, both of which finally brought the life scientist and engineer into communication and constructive interaction, albeit with considerable skepticism on both sides about the consequences. One was the recruitment, by the more perceptive biotech firms, of biochemical and chemical engineering talent from the food, drug, and chemical industry to organize and build the essential process research and development and manufacturing capabilities needed to carry the new product from the laboratory into the marketplace. The other (available to ventures with promising products and good bioscience, but limited financial resources) was to seek collaboration (via acquisition, joint-venturing, or product-licensing) with a major pharmaceutical or chemical manufacturer possessing the basic engineering skills and resources required for commercialization. Those start-ups unable or unwilling to pursue either of these two courses were, the record will show, doomed to untimely demise, proving that two geneticists and a brace of white rats do not constitute a business.

During this same period, the major chemical and pharmaceutical manufacturers were hardly indifferent to the biotech revolution. They were both initiating their own genetic engineering programs, and building working relationships with biotech start-ups in order to establish a proprietary position in the more attractive new bioproducts. But even more importantly, however, these companies, because of their long-standing experience and expertise in chemical and biochemical engineering, marshalled their process engineering staffs early on to begin to grapple with the problems of production and scale-up of the emerging new bioproducts. As a consequence, biotechnology developments in these companies have benefited from active engineering participation from their inception, and have progressed quite rapidly to successful commercialization.

This survival-driven liaison between life scientist and engineer has, over the past few years, been a fascinating development in which I, for one, have had the good fortune to participate. What commenced as a furtive association between a proud, confident, and xenophobic pure science discipline skeptical about what engineers had to contribute to their labors, and an engineering discipline mystified by the arcane concepts and practices of the biological scientists, and unsure of its ability to solve their problems, has now begun to evolve into a mutually supportive, mutually respectful, and increasingly productive union. Fortunately, many of the established tools and techniques of chemical and biochemical engineering that have been successfully employed in the design of manufacturing processes for traditional pharmaceutical chemicals and fermentation products are proving to be adaptable to the manufacture of the newer bioproducts. This is evidenced, for example, by the successful commercialization of human insulin generated by recombinant microorganisms (an Eli Lilly/Genentech collaboration), the imminent commercial introduction of genetically-engineered bovine growth hormone (by Monsanto), and the recent registration of recombinant human growth hormone by Genentech.

Possibly an even more important consequence of the marriage of life science and engineering is the increasing awareness by engineers that many of the sophisticated life-science laboratory practices employed to grow and manipulate living cells, and to assay and isolate biological substances, contain the seeds of novel and useful means for large-scale production and purification of these products. Conversely, microbiologists, cell physiologists, molecular biologists, and biochemists are beginning to recognize that, by exposing their bioengineering colleagues early on to the special properties and characteristics of the cells they are attempting to manipulate, and of the bioproducts they hope to produce, they can obtain advice, guidance, and innovative suggestions about their procedures which not only are improving their chances of success in the laboratory, but are greatly expediting and simplifying their ultimate adaptation to large-scale production.

As might be expected, this vital cross-disciplinary collaboration between life scientist and engineer, on which the success and survival of the emerging biotechnology industry will depend, is today largely limited to that small number of corporations which have a vested commercial interest in bioproducts with relatively near-term market potential. These, in the main, are human and animal health-care products for diagnostic, prophylactic, and therapeutic applications. The U.S. companies involved are the larger drug and chemical companies (e.g., Merck, Lilly, Monsanto, DuPont), and the well-established genetic engineering firms (e.g., Genentech, Cetus, Biogen). These organizations are establishing highly competent and innovative interdisciplinary bioprocess R&D teams, comprising engineering-oriented life scientists and life science-oriented engineers, well-equipped to serve their near-term needs.

Regrettably, however, this important new bioprocess R&D resource is being nurtured and developed within a quite small and secretive segment of the rapidly expanding biotechnology industry, and can hardly be expected to supply the talent and know-how required to meet the process R&D needs of the succeeding generations of biotechnology-based businesses, that is, those directed to biologically-derived nutritional and agricultural products, industrial chemicals, and fuels.

Moreover, our universities are today unable to educate and supply to the industrial sector even traditionally-trained biochemical engineers in numbers adequate to satisfy the anticipated requirements of the biotechnology industry. Of even greater concern is the fact that few if any of our most prestigious universities are yet able to persuade their life science and engineering faculties to cooperate in the creation of imaginative teaching and research programs aimed at producing the unique species of life science-oriented engineer (or engineering-oriented life scientist) which is so desperately needed by the modern biotechnology industry.

Thus, we are faced with a crisis of shortage of properly trained and qualified life science and engineering personnel equipped to transform our newly-found powers of genetic manipulation into industrially useful and profitable products and processes. Without an early solution to this problem, the momentum which has characterized the biotechnology revolution of the past decade is likely to be lost, and with it, many of the hopes of societal benefit from its discoveries.

KEEPING THE BIOTECHNOLOGY SHIP ON COURSE: THE ENGINEERING CHALLENGE OF THE COMING DECADE

The successful commercialization of the first-generation bioproducts based on modern genetic engineering has been, as noted earlier, attributable to fruitful collaboration between life scientists and chemical and biochemical engineers, who have welded established biological laboratory practices with available chemical/biochemical process technology to yield acceptable manufacturing processes for these relatively low-volume, high-unit-value products. Now receiving increasing attention, however, is an expanding family of human/animal health-care products (exemplified by such substances as tissue plasminogen activator, therapeutic monoclonal antibodies, and viral surface antigen vaccines) which require the use of living mammalian cells for their synthesis, and must be recovered in exceedingly high purity. Technology for the industrial-scale generation and purification of such products is today at best rudimentary, and in many cases non-existent. Succeeding generations of bioproducts—agrichemicals such as herbicides, pesticides, and plant growth regulators; nutritional chemicals such as amino acids and vitamins; food additives such as sweeteners, flavors, fragrances, single-cell protein, fats and oils; industrial organic chemicals such as solvents, pharmaceutical intermediates, plastics, plasticizers, and lubricants; and liquid and gaseous fuels—will require development of bioreactors of unprecedented size and efficiency, of product recovery techniques of extraordinary selectivity and energy-economy, and of novel waste-reprocessing and pollution abatement systems dwarfing in capacity and sophistication those in use at present. While some of these emerging needs will probably be met by improvement of (or imaginative combination of) existing process technologies, most will require development of entirely new devices and systems calling on the creative genius of the life scientist and the engineer in concert.

The industrial progress of biotechnology will obviously be paced by the rate at which we can solve these formidable problems in bioprocess engineering. Neither our existing educational institutions, nor our current reserve of scientific and engineering talent, are equipped to perform this task, let alone to provide the numbers of qualified experts which the industry must employ to accomplish its objectives. Today, the industrial demand for bioengineers is actually having a negative impact on their future availability, as biotechnology companies compete for and lure away from academia those individuals best qualified to train the next generation of specialists in the field.

Nor can we expect the biotechnology industry itself to be the nation's source of the talent-pool in bioprocess engineering, or of a publicly-accessible knowledge base in this important technology. The dynamics of a free-market economy militate against sharing of proprietary knowledge, and unimpeded movement of specialists between competing organizations.

As our industrial, academic, and governmental planners and administrators ponder this problem, events beyond our control, but within our awareness, have begun to threaten our early leadership in the promising field, and to raise the specter of America's reversion to a second-or third-rate player in the global biotechnology game. Our technologically-advanced global trading partners, principally Japan and West Germany, are now marshalling their intellectual, human, and financial resources very aggressively to build their national competence in engineering for biotechnology, and to put into place the appropriate multidisciplinary institutions committed to meeting industry's needs for the long term. They are achieving this objective rapidly because of their awareness of the essentiality of interdisciplinary team-effort to progress, and a strong cultural commitment to national dominance in new technology rather than limited benefit to the few. In both of these countries, the relationship between industry and academia has been much closer and more supportive than that in the United States, as has the relationship between the federal government and the university system, and between government and industry. Hence, it has been relatively easy for these governments to orchestrate the establishment of research, development, and training programs and institutions staffed by government, industrial, and academic technologists and managers, whose goals and objectives are well-defined, and enthusiastically supported by all participants.

In this country, however, conditions and attitudes are far from conducive to the establishment of such joint public and private sector team projects focused on new technology exploitation. Firstly, our universities—the training grounds for our scientists, engineers, and managers—are fundamentally averse to the concept of team effort (particularly when disparate disciplines are needed) as a vehicle for research or education. Historically and traditionally, the path to reward in academia is individual accomplishment. Co-authorship among peers in academic research is regarded as weakness rather than strength. Crossing of departmental lines by university faculty is, more often than not, considered a sign of ineptitude. In contrast, the merits of interdisciplinary team effort have been persuasively proven in industrial R&D; hardly any successful R&D manager would fail to sing its praises. Unfortunately, however, technical accomplishments of industry seldom become part of the public record, nor do the managers of such enterprises often share their experiences with others. Secrecy is the name of the corporate game. This very concern with protection of proprietary rights is also a major obstacle to industry collaboration with academia in discovery areas of technology. The academic commitment to free exchange of information is in direct conflict with industry's compulsion to restrict its disclosure.

Possibly the greatest barrier to effective industry/university/government collaboration in frontier areas of technology is the adversarial relationship which exists between industry and government, under the guise of the necessity to "protect our free enterprise system." A statutory structure which attributes conspiracy in restraint of trade to virtually any collaborative activity between industries or companies aimed at sharing resources and sharing the fruits of that association, cannot deal effectively with a situation (as now exists in biotechnology) where (1) the essential resources are in short supply, (2) the costs of resource and technology development are beyond the reach of any one industry, and (3) the risk of inaction or delay is the loss of a critical national asset.

Fortunately, however, new forces are at work within our educational, governmental, and industrial institutions seeking to mitigate these obstacles. Many of our most distinguished academic institutions are attempting to establish and build (in consort with industry and state or local governments) centers-of-excellence in biotechnology, within which faculty representing all the critical scientific and engineering disciplines can work together in research and teaching at the frontiers of the field. Many of our large corporations are today sponsoring in our major universities broadly based, long-range interdisciplinary research programs in biotechnology and bioengineering. And, within the past year, the federal government has undertaken to seed centers for interdisciplinary engineering research in "leading-edge" areas of technology (bioprocess engineering being one) at universities possessing the appropriate range of basic and applied science and engineering capabilities. These Engineering Research Centers, sponsored by the National Science Foundation, are intended to draw upon industrial participation and partial support from their inception, and to become entirely dependent upon industry for their support after five years. This concept of "need-focused" centers cutting across traditional university departmental lines is a significant departure from usual federal support practice. Whether the tactic will achieve its objectives, only time will tell.

Hopefully, these diverse efforts now underway to meet the challenge of commercializing biotechnology may lead to a coordinated, soundly managed national program of bioprocess engineering research, development, and education. Achievement of this objective will clearly require some major changes in attitude, reordering of priorities, and new accommodations between educators, industrialists, and the federal bureaucracy unlike any we have seen before. If we fail, we shall probably not have a second chance. The next decade will be a trying one for biotechnology, and a fateful one for the nation.

SUMMARY

The industrial reduction-to-practice of the discoveries of modern biological science is the principal task of bioengineering. Only recently have life scientists begun to recognize this fact, and only recently has the fledgling biotechnology industry begun to enlist the aid of the engineering community in development of commercial-scale production facilities for its products. The technology base of modern chemical, biochemical, and bioprocess engineering, while suitable for the design of manufacturing plants for certain bioproducts, is inadequate for production of most human/animal health-care biologicals, and seriously deficient for the large-scale production of second generation bioproducts such as agrichemicals, nutritional chemicals, industrial chemicals, and fuels. Moreover, the supply of biochemical and bioprocess engineers to be trained by our universities in the near future falls far short of the projected needs for these skills by the biotechnology industry. These limitations must be overcome by establishment of a nation-wide interdisciplinary research, development, and educational program in biochemical/bioprocess engineering, involving collaborative interaction of life scientists and engineers, and liberally staffed and funded jointly by government, industry, and academia. If we fail to respond promptly and aggressively to this challenge, our opportunity for international domination of a vital new technology which America largely originated will be lost to other nations with greater commitment to national goals.

REFERENCES FOR FURTHER READING

Bloch, E. (1986). Basic research and economic health: the coming challenge. *Science* 232: 595-599.

Committee on Science, Engineering, and Public Policy, National Academy of Sciences, National Academy of Engineering, and Institute of Medicine (1984). Research briefings 1984 for the Office of Science and Technology Policy, National Science Foundation and selected federal departments and agencies. *Report of the Research Briefing Panel on Chemical and Process Engineering for Biotechnology*. National Academy Press, Washington, D.C.

Dubinskas, F.A. (1985). The Culture chasm: scientists and managers in genetic-engineering firms. *Technology Review* 88: 24-74.

Giamatti, A.B. (1982). The university, industry, and cooperative research. *Science* 218: 1278-1280.

Hill, T.H. (1985). Rethinking our approach to science and technology policy. *Technology Review*, April, 1985.

Hochhauser, S.J. (1983). Bringing biotechnology to market. *High Technology* 55-60 (February, 1983).

Humphrey, A.H. (1984). Commercializing biotechnology: challenge to the chemical engineer. *Chem. Eng. Prog.* 80 (12): 7-12.

Keyworth, G.A. (1985). Science and technology policy: The next four years. *Technology Review*, February/March 1985.

Michaels, A.S. (1982). Frontiers of chemical engineering. *Chem. Eng. Commun.* 17: 99-106.

Michaels, A.S. (1984). The impact of genetic engineering. *Chem. Eng. Prog.* 80 (4): 9-15.

Michaels, A.S. (1984). Adapting modern biology to industrial practice. *Chem. Eng. Prog.* 80 (6): 19-25.

National Academy of Engineering. (1982). *Genetic Engineering and the Engineer—A Symposium*. National Academy Press, Washington, D.C.

National Research Council, Board on Chemical Sciences and Technology: Committee on Chemical Engineering Frontiers: Research Needs and Opportunities: *Report of the Panel on Biochemical and Biomedical Engineering*, March, 1986 (unpublished privileged document) National Materials Advisory Board, Commission on Engineering and Technical Systems, National Research Council (1986). *Report of the Committee on Bioprocessing for the Energy-Efficient Production of Chemicals*. Publ. No. NMAB-428, National Academy Press, Washington, D.C. (April, 1986).

Reich, R.B. (1983). The next American frontier. *The Atlantic Monthly* 97-108 (April, 1983).

U.S. Department of Commerce, International Trade Administration. (1984). High Technology Industries (1984): *Profiles and Outlooks—Biotechnology*. Washington, D.C.

U.S. Congress, Office of Technology Assessment. (1984). *Commercial Biotechnology—An International Analysis*, U.S. Government Printing Office, Washington, D.C.

Webber, D. (1985). Consolidation begins for biotechnology firms. *Chemical & Engineering News* 25-60 (Nov. 18, 1985).

PART IV
GLOBAL, UNIVERSITY AND INDUSTRY PERSPECTIVES

A European Perspective on Biotechnology

John Maddox
Editor, Nature
London, England

Whether there is a distinctively European perspective on biotechnology and its future development is for non-Europeans, rather than me, a European, to judge. But there are several respects in which European attitudes color perspectives on all kinds of issues.

For example, it must be confessed that Europeans have become an envious lot. We envy the United States for its capacity to get things done, its capacity to execute bright ideas, turning them into commercial projects. And then we envy Japan for its capacity to do the same things, often faster, usually more cheaply, and also with a flair for selling them in the world's markets that outstrips that of the United States.

Europeans are also more used to being disappointed. Over the years, but certainly since the Second World War, there have been several occasions when European governments or companies have embarked on technical projects which they had good reason to believe would establish a commanding lead in some part of some field, only to find that their hopes were dashed. Britain, for example, probably had the most technically advanced computer in the world in 1947. A few years later, there was the world's first commercial jet aircraft, the Comet, that might have swept the world had it not kept falling out of the sky. Tales like this can be repeated in France, West Germany and most other West European countries. Whatever happened to the Italian chemical industry that founded the economic miracle in Italy in the decade from 1958?

This long record of disappointment is not, of course, unrelieved. The European pharmaceutical industry, for example, is a substantial third of the world's total; it can be so only because it is technically as advanced as any. And there's a long list of other European companies which are technically strong and commercially successful. It's also the case, in my opinion, that Europe has learned, the hard way if you like, why there have been so many disappointments: cultural snobbery, the belief that ideas arising in the continent with the longest recorded history of human occupation must be intrinsically and in some sense better than those arising elsewhere, has now largely gone; stemming from that fault is Europe's traditional view that there's no great need to match goods carefully to the needs of their intended customers, or to manufacture them for reliability and durability, but it'll be some time before that fault is cured; and then there's the trade-off Europeans always make between the discomforts of the change occasioned by technological innovation and the collective fondness for ancient institutions, universities and the like, where Europe may continue to pay a high price for keeping things the way they are.

With all this said, European habituation to disappointment does at least have the virtue of making us less impatient at the fact that projects flounder from time to time. We're resigned. Philosophical is the grander name for this state of mind. Maybe you'll think it's colored what I have to say about biotechnology.

REAL REVOLUTION

There's no doubt that the industry about which this conference is concerned amounts to a revolution in technology. It's important to be clear just why this is the case. We were reminded by last night's dinner speaker how it all began. By 1972, molecular biologists working at universities had realized that it is possible to manipulate naturally occurring strands of DNA and (with the help of reverse transcriptase) RNA as well. The legend of how Professor Paul Berg was persuaded of the potential, and the possible risks, of these developments (at a Gordon Conference, by graduate students) is now part of the folklore. There followed the moratorium on experiments with genetic manipulation, so that it was not until 1977 or thereabouts that people were free to embark on actual genetic manipulation.

You'll recall that this interval was used for two distinctive purposes. First, there was endless discussion about how (and even whether) genetic manipulation should be controlled. And there was endless speculation about the things that would be done when the moratorium was ended.

It's interesting to remember how optimistic that speculation turned out to be. The idea that it would be possible to use bacteria carrying genetically manipulated plasmids to manufacture any naturally occurring protein was taken for granted. Vaccines would be readily developed to match any identified virus or bacterium in the natural world. People talked about the possibility of grafting nitrogen-fixing genes from the plant-parasitic rhizobia in which they naturally occur into plants, which would then have the facility to make their own fertilizer from atmospheric nitrogen. It's ironical to remember that, during that period, there was a lively argument in Britain on the issue whether the transplantation of nitrogen-fixing genes would be in the national interest, given that the plants with the transplanted genes would grow successfully in the tropics. In the event, of course, it's turned out that the nitrogen-fixing genes are a complicated complex of genes, and that nobody would now think of transplanting them in the way I've described, at least without a great deal more information about the way in which they're regulated.

Nevertheless, I think it's fair to say that there were two substantial elements in the circumstances of the mid-1970s that justified the optimism of the times. First, there was every reason to expect that the manipulation of DNA by restriction enzymes would be far easier than the chemical manipulation of naturally occurring molecules; nothing that has happened since suggests that this assumption was incorrect, even though the chemists are now much cleverer than they were then. Second, there was an almost unparalleled identity of interest between the academic scientists and those in industry. Both, for example, need to know how genes are regulated in natural organisms. This common cause persists although, as you'll hear, that circumstance may not be the universal blessing that you suppose it to be.

Nothing that's happened since the ending of the moratorium suggests that the ultimate promise of biotechnology is any less exciting than people then expected it to be. It is a real revolution of technique.

So much has been widely recognized. Every government with technological pretensions has a biotechnology program of some kind, intended to support its domestic industry with basic research or even, sometimes, with outright subsidies. Every government in Western Europe has such a program. So, too, does the European Commission, the executive of the European Community. The Soviet Union has a program. So does India. So, too, does China. What these governments hope is that their programs will enable their industrial companies to capture a share of what may be a very large market. Some, of course, hope to capture a commanding lead. For others, it is sufficient that there is a chance of winning a proportionately small piece of the pie which, because the pie may be very large, will be an invaluable asset. And sometimes, the ambitions of the sponsoring governments are outrageously ambitious. Two years ago, a distinguished U.S. scientist and I spent a frustrated evening trying to persuade two high officials of the government of India that they were mistaken in believing that investing in research in biotechnology would enable India, within two or three years, to begin exporting genetically engineered chemicals in competition with the world's pharmaceutical industries. The evening was frustrating because the persuasion failed.

HARD REALITY

One of the peculiar features of the present state of biotechnology is that the old optimism of the early days is sustained in the face of recent experience that's more than a little disappointing. A few examples will show you what I mean.

It was natural enough, fifteen years ago, to think that the manufacture of human insulin by the techniques of biotechnology would be a natural first step. There were strong reasons to suppose that human insulin would be better in the treatment of diabetes mellitus than the pig insulin previously most widely used. It might even be cheaper. And human insulin was indeed the first therapeutic product to come off the biotechnology production lines. But then a smallish pharmaceutical company in Denmark found it could also produce human insulin by modifying the pig version by traditional chemical techniques. As I understand it, both processes are now used in parallel. What these developments show, however, is that biotechnology will not necessarily sweep everything else before it.

The case of interferon is more complicated. It has been maybe ten years since it seemed as if a plentiful supply of this naturally-occurring biochemical, firmly implicated in the body's natural defenses against infection, would be a pharmaceutical product of immense value, and one that almost by definition would be free from harmful side-effects, while there were also good reasons to hope that a product which was plainly the central element in the immune system might also have a role to play in the treatment of cancer. By the beginning of this decade, several of the many biotechnology companies that had sprung into being were hard at work on the manufacture of interferon by means of engineered plasmids in bacteria.

But several previously unsuspected obstacles quickly came to light. There are three types of interferon (alpha, beta, and gamma) with very different properties. And there are no fewer than thirteen genes making similar (but not quite identical) molecules of the alpha type, the evolutionary reasons for which are not understood. Then, perhaps inevitably, it turned out that there are side-effects, chiefly because the interferons are not, as they had been thought, simply anti-viral substances but modulators of the immune response which, among other things, may regulate the growth of cells. The result is that there's now an ample supply of genetically-engineered interferon, much of it metaphorically kept in cold-store. Time will tell whether it will have other than marginal uses in therapy, perhaps in conjunction with other materials. And the chances are that much the same fate will befall some of the other naturally-occurring materials that have more recently excited interest, the interleukins and even tumor necrosis factor (TNF).

The moral is that the biotechnology is every bit as powerful, even easy in a relative way, as the enthusiasts have been saying, but that the biology is more difficult than people have been willing to allow. Exactly the same applies to the nitrogen fixation genes. In spite of all the pace of discovery in molecular biology in the past two decades, the plain fact remains that it has served also to sharpen our appreciation of the ignorance remaining.

REVOLUTION BY DEGREES

Fifteen years is a substantial length of time, especially for a process called a revolution. If much more time passes, it may no longer be permissible to use the term revolution. So let me put the arguments in extenuation of biotechnology. There's the promise that lies ahead.

This year, there will almost certainly be trials of a malaria vaccine, probably based on a combination of the Melbourne and New York antigens. It would be foolish, in the light of what's been learned in the past fifteen years, to expect that this project will produce a workable vaccine at the first attempt. But there is a high chance that enough is now known of the antigenicity of the malaria parasite for it to be clear that the project will work at some early stage, that there will be an effective malaria vaccine in the next decade, say. When that time comes, who will dare say that biotechnology has been a disappointment?

Nearer ahead, there's the prospect that the diagnosis of certain genetic defects *in utero* by strictly genetic techniques is literally at hand. Amniocentesis has been widely used for twenty years, of course, chiefly for the diagnosis of congenital malformations (Down's syndrome, for example, where visual inspection of chromosomes will suffice) as well as for the recognition of genetic defects which are recognizable by the products produced within the amniotic fluid by the presence (or absence) of known enzymes (which is how phenylketonuria can be recognized in advance). But the chance of recognizing genetic defects such as cystic fibrosis depends entirely on being able somehow to inspect the DNA of the unborn fetus, which in turn requires a genetic probe as well as a sample of DNA (perhaps more conveniently from the placenta than from the cells of the amniotic fluid). Probes that will do this are now beginning to accumulate. There may be dozens of them in a few years. Whole classes of genetic defects may become avoidable.

Then there's the disease called AIDS. Biotechnology hasn't cured AIDS, nor will it. But it's unthinkable that so much could have been learned in so little time about this novel disease were it not for the use of the techniques of cloning and sequence analysis which are the basis of biotechnology. And the same techniques have led directly to the diagnostic techniques from which will spring means of controlling the spread of the disease and perhaps, eventually, better means of dealing with the symptoms. That's something to be proud of.

So there have been many disappointments in the past fifteen years, but the promise is now much more solid than it was fifteen years ago. It's a revolution right enough, but a slow revolution.

The surprise is that we're surprised by that. It's usually like that with major innovations of technology. I remember, in 1960, that the then British government, together with the Institute of Production Engineers and a number of British trades unions, held a major conference in Britain on the impact of automation. The argument was simple. Robots based on computers are about to arrive, so we must plan immediately to deal with the problems that will arise, the people displaced from jobs, the leisure that everybody will have to fill. The conference, as I recall, was told a little about the outstanding technical problems. Robots needed slightly better sensors; the computers could with this advantage be a little faster, and more compact, and there were certain conceptual difficulties in deciding what to tell the computers. But these were practical problems that would be easily solved. In reality, as we now all know, they haven't yet been solved, and this particular revolution has taken a quite different course.

The moral is that important changes of technique are necessarily slow. We mustn't expect otherwise.

THE LEARNING CURVE

It's also important to remember that, if there have been disappointments in the past fifteen years, we also learned a lot. In Europe, for example, a little to our surprise, we've discovered that Europeans, too, can become entrepreneurs. Chiefly under the influence of what was happening in the United States in the 1970s, Europe has become a hunting ground for venture capitalists. It's rare that a person with a good idea, a plan for setting up an independent company, and an ambition to become rich, will fail to find people willing to talk about the terms on which they will back him with resources. Moreover, the phenomenon is not confined to biotechnology; the venture capitalists are working other fields of new technology. In the long run, the result of all this cannot fail to be beneficial to countries such as Britain.

Of necessity, many of the new entrepreneurs are academics, which raises other questions. As it happens, the excitements about biotechnology coincide with the anxiety of governments of all kinds (including that of the Soviet Union) that academic science should be a more effective instrument for the creation of prosperity. One result has been that academics and their universities have been caught up in the process of industrial innovation, in no field more fully than biotechnology. I happen to believe that the experience, largely in the United States, in the past five years has demonstrated that it is possible to avoid some of the obvious dangers that there will be insupportable conflicts of interest. But I am also concerned that a large proportion of the academics working in this field are now very closely tied in with commercial goals.

That could be a dangerous development not because academics have no business contributing to commercial success, but because too much preoccupation with industry's affairs can be distracting. In the long run, this could harm the industry itself; the example of Interferon shows how great is the need for a deeper understanding. There's also the danger that, if too many talented people in this field are occupied emotionally as well as intellectually with industry, they will not be tempted energetically to explore the other fields of science that lie ahead of them. It could even be that the still-persisting identity of interest between academic and industrial science is a sign that the distortion of the natural pattern of the development of science is already apparent.

There's another problem to which the universities will soon have to give their attention, that of providing the supply of skilled people on whom the revolution must depend. I think we've consistently and grossly underestimated the numbers of people who will be needed, and the degree of the skill they will need if they are to be effective in, say, the management of immediate problems such as the application of gene probe analysis for genetic defect.

Finally, there's the problem of regulation. The moratorium was a crude beginning. My own view now is that the period since it ended has shown as clearly as anybody could ask that the scientific community deals responsibly with the need to conduct laboratory experiments safely. At the beginning, it seemed only prudent to fall in with the idea of a moratorium, unprecedented though it was. As things have turned out, I think it is time that responsibility for deciding what laboratory experiments should be carried out should be passed back to the laboratories.

The regulation of the release of genetically engineered organisms (plants, bacteria, and viruses and perhaps, eventually, insects) is a different issue. Elsewhere, we cannot understand why the United States is making such heavy weather of this matter; no doubt it has a lot to do with the Constitution and with people's access to the courts. But the simple scientific solution is that there should be some careful demonstrations, supported by public funds of what happens when genetically engineered organisms are released into the environment. To begin with, of course, they would be labelled not by the kinds of genes that people might worry about, but by harmless marks. The question that needs testing is whether we are right in our expectation that the organisms that survived are those which have found their way into the natural environment as a consequence of natural selection. Then it should be much easier to tell what happens when Mr. Jeremy Rifkin takes yet another court case against a government agency.

This is an important issue, and one that needs urgently to be solved. The scientific community as a whole, not just the biotechnologists, need to battle with it. For by now it's plain that the revolution called biotechnology is not only a little slower than some people had been hoping, but also that it'll be a long one, a bit like that between the beginning of the eighteenth century and the beginning of the nineteenth century that saw the practical application of the steam engine. There's no reason why we should accept that it should be longer than is necessary.

The Status and Future of Biotechnology in Developing Countries

Yongyuth Yuthavong
National Center for Genetic Engineering and Biotechnology and Department of Biochemistry, Mahidol University, Bangkok, Thailand

INTRODUCTION

The general excitement on the potentials of biotechnology, focusing on novel genetic engineering techniques, is not lost on the developing world. Attracted by the prospects of a technology with wide applications, accessible in many respects even with modest capital, the resource-rich and resource-poor countries of the developing world alike are eager to have a share of this technology. The resource-rich countries hope to turn their resources to value-added products, while the resource-poor countries hope to develop their resources through biotechnology. Needless to say, the obstacles are tremendous, one of the most important being lack of skilled manpower. However, compared to other technologies, biotechnology appears to be unique in satisfying the criteria of being both a "high" and an "appropriate" technology, appealing to both ends of the policy spectrum in the developing world. It seems possible to overcome the foreseen obstacles to a certain extent provided the determination and sufficient input. However, as those concerned with policy decisions are beginning to realize, the benefits of biotechnology in general may require medium to long-term input in order to be realized, and a sustained effort of a magnitude over a minimum threshold is necessary.

THE GLOBAL PERSPECTIVE

The magnitude of problems besetting the developing world can be appreciated from a glance at some statistics. Table 1, for example, gives a comparison of the situation in agriculture and food between the developing and developed countries (FAO, 1981). Biotechnology has made, and will continue to make, great impacts on agricultural and food production.

The outlook on world agricultural production has changed considerably recently, owing significantly to the advances of biotechnology, resulting in favorable increases from developing and developed countries alike. However, the general decline in prices of agricultural commodities has added another dimension to the problems. Table 2 gives an estimate of the number of people in the developing world suffering from important infectious diseases (Warren, 1985), the alleviation of which depends to a great extent on vaccines and diagnostics developed from new biotechnology. From these selective indicators, it is obvious that biotechnology can contribute significantly to the problems of the developing countries.

Table 1. Comparison of the situation in agriculture and food between developing and developed countries.

	Developing Countries	Developed Countries
Percent of world population	67	33
Percent of world agricultural production	38	62
Production per agricultural worker ($,1975)	550	5220
Arable land per agricultural worker (ha)	1.3	8.9
Fertilizer use (kg/ha) of agricultural land	9	40
Total daily food consumption (calories)	2180	3315

Table 2. Estimate of the number of people in the developing world suffering from important infectious diseases, 1977-1978.

Infection	Deaths (thousands/Year)	Disease (thousands of cases/Year)
Malaria	1200	150,000
Schistosomiasis	500-1000	20,000
Tuberculosis	400	7,000
Hookworm	50-60	1,500
Ascariasis	20	1,000

Source: WHO and Special Program for Research and Training in Tropical Diseases.

A crucial question in the use of biotechnology in the developing world is its endogenous capability and potentials for development. Many developing countries have, through the centuries, accumulated traditional wisdom on farming techniques, fermentation methods, etc., which can serve as a base for development of new biotechnology. However, the bulk of new biotechnology presently making an impact on the industrial world is science-based and depends on existing industrial infrastructure. While a comprehensive survey of the scientific and industrial infrastructure relevant to the development of biotechnology in various countries of the developing world is still lacking, a glimpse can be obtained, for example, from Table 3 which shows basic facts about disparities in science and technology between developed and developing countries (UN, 1979). Obviously, international cooperation is needed for promotion of biotechnology in the developing world. Since the developing world represents a large potential market and carries vast bioresources, this promotion should be in the interest of the developed and developing world alike.

Table 3. Some quick facts about disparities in science and technology between developed and developing countries.

Distribution of World Research and Development Expenditures in 1973	US Dollars billions	Percentage of world total
World total	96.4	100.0
Developing countries	2.8	2.9
Developed countries	93.6	97.1
Distribution Researchers (Research and Development Scientists and Engineers) in 1973	Thousands	Percentage of world total
World Total	2279	100.0
Developing countries	288	12.6
Developed countries	1990	87.4

Source: UN (1979).

INTERNATIONAL AND REGIONAL STATUS OF BIOTECHNOLOGY COOPERATION

International and regional attempts to harness life sciences for the benefit of agriculture, health, and the economy of the developing world in general were launched even before the impact of genetic engineering started to be felt. The International Rice Research Institute (IRRI) in the

Philippines, formally established in 1960, is an early example of success of such international cooperation, giving to the world the IR8 and a host of semi-dwarf rices which now account significantly for the present high world productivity in rice. Other institutions established later, including the International Maize and Wheat Improvement Center (CIMMYT) in Mexico, the International Laboratory for Research on Animal Diseases (ILRAD) in Kenya, together with the IRRI, are now supported by the Consultative Group on International Agricultural Research (CGIAR), in turn sponsored by the Food and Agriculture Organization (FAO), the World Bank, UN Development Programme (UNDP), various governments and funding agencies. The CGIAR is often cited as a successful model of international cooperation in research and training for a specific mission, i.e., to increase food production in the developing world. Another example of global cooperation for a specific mission, to which biotechnology is making a significant contribution, is the UNDP/World Bank/World Health Organization (WHO) Special Programme for Research and Training in Tropical Diseases. Started in the mid-seventies, the Special Programme aims at the alleviation of six major tropical diseases mainly predominant in the developing world. The major tropical diseases are also the main concern of other non-governmental organizations, e.g., the Rockefeller Foundation Programme on Great Neglected Diseases of Mankind. A number of new drugs and vaccines are being developed through these programmes.

A few efforts on the international scale have focused on biotechnology as the major tool towards development of the Third World countries. The effort by UNIDO to set up the International Center for Genetic Engineering and Biotechnology (ICGEB) is the most extensive one with original participation from both developing and some developed countries, although it has unfortunately suffered from the disinterest of the major countries. These problems are partly due to the expectations and fears concerning international transfer and development of this powerful technology, much of which is proprietary and involves major trade and industrial interests. Nevertheless, the establishment of ICGEB has sensitized many developing countries to the potential benefits of biotechnology, and the Center itself, with the two components in Trieste (Italy) and New Delhi (India), may yet make significant impact on the developing countries, depending on acquisition of a high calibre staff and a well-chosen programme of activity. An appraisal of the UNIDO effort to set up the ICGEB has been recently given (Zimmerman, 1984 a,b).

Other efforts in international cooperation on biotechnology for the benefit of developing countries include the agreement between the United Nations University and Venezuela to set up the International Institute for

Biotechnology. A series of conferences on Global Impacts of Applied Microbiology (GIAM) have created awareness in a major aspect of biotechnology since the sixties. The International Organization for Biotechnology and Bioengineering, a non-profit organization created in 1968 by individual scientists, has been responsible for various training courses and international symposia with emphasis on the developing countries. Some developed countries, e.g., USA (BOSTID, 1982) and the Netherlands (van Hemert *et al.*, 1983), have devoted interest to the potentials of biotechnology in developing countries and the role of international cooperation.

In southeast Asia, cooperation in science and technology among the six-member countries of the Association of Southeast Asian Nations (ASEAN) have, in the last decade, given rise to many research and training programmes in biotechnology related areas. A specific programme for biotechnology has been launched recently, with cooperation from Australia and the EEC. The major areas of emphasis will be development of bioactive agents from plants, plant tissue culture, rhizobium biology, diagnostic agents, industrial enzymes and pilot plant design, and bioreactor control. ASEAN is enormously rich in natural resources, being the only region in the world which is both a net exporter of food (from Thailand) and of energy (from Brunei, Indonesia, and Malaysia). It is a region of 280 million people and an estimated $220 billion market. The development of biotechnology in ASEAN has substantial implications for the other industrially developed countries, not only in terms of market for consumer biotechnology products, but also in terms of possibility for overseas investment and joint venture and transfer of technology for local production. The decline in commodity prices are increasingly influencing the policies of ASEAN governments towards increasing the value-added of their products through biotechnology.

STATUS OF BIOTECHNOLOGY IN SOME DEVELOPING COUNTRIES

The status of biotechnology in some developing countries is considered here in brief in order to obtain a picture for comparison with the more well-known status of biotechnology in more industrially developed countries. The countries considered here are drawn from the ASEAN region, representing a middle-stage and moderate to fast-developing status. The status of biotechnology in Africa, Latin America, and other countries has been discussed elsewhere (e.g., van Hemert *et al.*, 1983).

THAILAND

Three outstanding reasons for Thailand's active interest in biotechnology are: (1) it is a fertile country with a vast pool of unused natural bioresources, (2) for a developing country of her size, Thailand has a relatively large number of qualified personnel for biotechnological research and development, and (3) it has a relatively sound economic status so as to ensure returns in judiciously chosen biotechnology investments. Development of biotechnology in Thailand can be viewed in a historical perspective as arising naturally from the growth of both supply and demand sides of the technology. From the supply side, the last two decades in particular have seen a remarkable growth and maturity of basic life sciences in Thailand, linked with the emergence of a critical mass of scientists trained in developed countries and the establishment of research-based graduate schools. The latter development has been responsible for the presence of substantial endogenous research capability. From the demand side, problems in agriculture and rapidly expanding agro-industry, health and the environment have made it necessary to utilize biotechnology at various levels of complexity. Adequate linkage is, however, still presently lacking between the supply side, mainly located in the university sector, and the demand side, mainly located in the private and the government sector. This lack of linkage still has to be remedied, through government policy and other measures to encourage the utilization of endogenous resources for research, development, and technology transfer.

The infrastructure for biotechnology in Thailand has been reviewed (Yuthavong *et al.*, 1984). Presently, Thailand has approximately 60,000 scientists and engineers with degree-level education. Of these, it has been estimated that the number of scientists and engineers, at M.Sc. level and above, active in R&D biotechnology is approximately 400. There are approximately 20 institutions, mainly in universities, engaged in R&D in biotechnology. Most of these are in the Bangkok metropolitan area. At least five universities are offering courses and training at graduate or undergraduate levels in biotechnology. Although the present strength is still rather low, it is likely to grow rapidly. One factor that has helped in strengthening biotechnology in Thailand is the establishment of the National Center for Genetic Engineering and Biotechnology in 1983. The Center formulates policy and plans on biotechnology, supports R&D activities in designated institutions and serves to link the institutions with the private sector. Presently, the Center has four affiliated laboratories and R&D projects in eight institutions. On-going projects range from various aspects of plant tissue culture, animal embryo technology, genetic engineering of microbial larvicides, and bioconversion of cassava starch. A report on some aspects of development of the Center has been prepared with input from a U.S. advisory group (MOSTE and BOSTID, 1986).

The interests of bioindustries in Thailand related to biotechnology include the areas of amino acid production for feed, cassava starch modification, hybrid seed production, commercial plant propagation through tissue culture, secondary production of antibiotics, and animal vaccines. The private sector mostly acquires technology through import, and it is hoped that future supply of local biotechnology expertise will help to a far greater extent than present in the choice, acquisition, and development of the imported technologies.

Another recent boost for biotechnology, as well as for other frontier technologies in Thailand, is the launching of the Science and Technology Development Project, through the U.S.-Thai bilateral government cooperation programme. The Science and Technology Development Board is chaired by the Deputy Prime Minister and comprises members from relevant government departments and representatives from the private sector, including the Association of Thai Industries and the Thai Board of Trade. The activities to be implemented include provision of R&D grants in biotechnology, provision of grants and soft loans for bioindustrial development, and promotion of linkage between the universities and the private sector. In order to ensure successful development of bioindustries and other science-based industries, the Thai government is in the process of reforming investment, tax laws, and regulations and of promoting R&D through tax incentives and other means.

SINGAPORE

Singapore's strong base in international trade provides a springboard for its development in biotechnology. The scientific infrastructure is provided by the National University of Singapore, which operates various programmes relevant to biotechnology at the undergraduate and graduate levels. The proximity of the university to the recently established Science Park should help in strengthening the link between university and industry. In addition, the proposed Institute of Molecular and Cell Biology, planned to become the center of excellence in the region with staff drawn from international sources, will generate additional supply of manpower as well as R&D results. The Institute will focus research in the areas of microbial genetics and physiology, immunology, and plant cell biology.

A number of local and foreign companies have invested in biotechnology in Singapore. The products range from plants propagated by tissue culture, food products from fermentation and enzyme processes, 6-aminopenicillanic acid for antibiotic production, and other health-care products. The government has various schemes to help finance R&D activities and commercialization in the private sector. The Product Development Assistance Scheme was set up in 1979 to stimulate growth of local and applied research and product development, while the R&D Assistance Scheme was set up in 1981 for funding-directed research. Tax incentives are also given for biotechnology and other businesses of pioneer or expanding status.

INDONESIA

Indonesia has great interest in biotechnology development, both because of its vast reserve of natural bioresources and because of its problems in food production and other areas which can be alleviated by biotechnology. Because Indonesia presently still has a very small manpower in biotechnology, the main thrust of its national programme is the buildup of required manpower, both through graduate education abroad and in selected local institutions. The government has recently established the National Center for Biotechnology in Cibinong, and three other interuniversity centers at Bandung, Bogor, and Yogyakarta. The Bandung Center will put primary emphasis on industrial biotechnology, the Bogor Center on agricultural aspects, and the Yogyakarta Center on health aspects of biotechnology (Oei Ban Liang, 1986).

MAJOR PROBLEMS AND CONSTRAINTS IN THE DEVELOPMENT OF BIOTECHNOLOGY

Although many developing countries are now strongly motivated to develop their biotechnology, a number of formidable obstacles are in the way towards their objective. The most important problems are lack of trained manpower and the weak scientific and technological infrastructure. The weak industrial infrastructure prevents the absorption and development of most technologies imported for commercial purpose. The technologies imported by the private sector often thwart the local R&D attempts which are mostly concentrated in the universities and government sector, already poorly linked with the private sector. Lack of access to information, both technical and commercial, hampers the effort to keep abreast with rapid advances in the developed countries. While most of the problems and constraints must be solved by internal efforts of the developing countries themselves, some are issues which require international cooperation. Some issues, furthermore, involve potential conflicts between the developed and developing countries, and should be considered with a view towards common solutions.

MANPOWER

Like other areas of science and technology, building up manpower in biotechnology in a developing country will necessarily take time, of the order of, say, five years at least. Since biotechnology is interdisciplinary in nature, the required manpower needs to be distributed over a broad base, ranging from molecular biology to process engineering. It may be necessary for a small developing country to be selective in its approach of manpower buildup, putting emphasis on the areas where it has comparative advantage, perhaps in specific natural resources. Whatever selective approach is used, a careful balance of emphasis on the basic and applied aspects of biotechnology should be achieved so as to generate self-reliance in the long run. Since the conditions for work in developing countries are drastically different from those in developed countries, training of manpower for a developing country may, at least in part, be wisely accomplished in laboratories, pilot plants, etc., of another developing country relatively advanced in the particular aspects of required training. Such training locations are difficult to find in developing countries and may have to be established specially for this purpose through collaboration with developed countries.

Beyond training in basic aspects of biotechnology, the development of biotechnology manpower in developing countries will require extensive exchange of personnel and joint R&D with scientists in the developed countries. Joint projects of common interest to researchers of both sides can be easily defined. A main criterion should be active participation of researchers from the developing countries on equal basis, as too often in the past such "joint" projects tended to be merely extensions of the scope of the interest of the partners from the developed countries or, worse still, suffer from a suspicion of being a convenient way for the developed countries to gain access to the bioresources of the developing countries.

ENDOGENOUS R&D CAPABILITY

The biotechnology capability of a developing country can survive and grow only if there is a substantial endogenous R&D capability. This is true even for countries with an open economy and an explicit policy to rely on technology import. This is so because for effective technology transfer, the local capabilities to receive, adapt, and develop the technology must exist, and these capabilities stem not only from training but from R&D activities. Indeed, the R&D activities in the developing countries rarely lead to important novel products or processes, but serve mainly as an effective means to achieve capabilities in specific areas of the technology.

The choice of R&D projects to be done meaningfully in the developing countries has been the subject of much discussion from various angles. It has been frequently argued that the R&D should serve the demand of local industries, and not simply done to extend the expertise of the researchers, serving only the supply side. This argument should in general be correct, although the present state of bioindustries in developing countries is usually not advanced enough to warrant many R&D projects. Therefore, if the demand factor is used as the only criterion, too many repetitious projects of trouble-shooting type might ensue. Even the major R&D projects designed according to demand would mainly be of the type of "re-inventing the wheel." For example, projects dealing with starch bioconversion might have to deal with purification and utilization of γ-amylases and glucoamylases, refined technologies for which have already been acquired by the major industrial enzyme companies in developed countries. It is nevertheless important for the developing countries to do such demand-generated R&D, with the provision that the R&D activities interact with transfer of available technologies from abroad. It is also important that basic research in, say, molecular biology or microbial physiology is not neglected, since modern biotechnology stems directly from advances in the basic areas of life sciences. A good balance between the various areas of R&D should be achieved, in order that a developing country can advance to maturity in biotechnology on a firm basis.

Many developing countries need to improve their agricultural productions, both for internal consumption and for export to obtain foreign currency. Endogenous R&D in biotechnology is a necessary component for the improvement of many areas of agricultural production. Although R&D in biotechnology related to agricultural production is an important activity, the fall in prices makes it presently unattractive to work on improving the yields of many major crops, and represents a dilemma for the developing countries. Another important area for the developing countries is biotechnology related to health care. The emphasis of the developing countries in health care delivery is increasingly shifted towards the preventive and community aspects. This requires development of vaccines, diagnostics, and drugs which can be administered on a large scale at low cost. Many health problems of developing world, e.g., tropical infectious diseases, are not shared by the developed world, and the R&D need to be done at least in part in the endemic areas. No matter which specific areas in agriculture, health or industry the developing countries choose to put major emphasis on, the R&D activity should have a high quality by international standards. This requires continuous interaction with the international biotechnology communities.

TECHNOLOGY TRANSFER

Since the bulk of biotechnology is presently done in developed countries, technology transfer is a major aspect of concern for strengthening of biotechnology capabilities in the developing countries at the commercial level and in general. Developing countries, especially those which are industrializing, import substantial amounts of technology, biotechnology included, under contract. The contractual agreements are often restrictive with respect to technology transfer and diffusion. The restrictions, understandable from the commercial point of view, pose a problem for endogenous development of biotechnology: the local technology cannot compete with the imported, yet inaccessible, technology. However, technology transfer is in many respects in the interest of both the technology owners and receivers. The investment and operating costs, for example, could be substantially lowered with effective transfer, local adaptation, and development. A balance, therefore, needs to be struck between the various aspects of technology transfer.

Many developing countries have tried to regulate transfer of technology through contractual agreements, in order to implement national policies on technology flow. A more promotive measure to regulate technology flow is to have technology information available, and perhaps a "match maker" unit to advise prospective buyers on suitability of the contractual agreements. Another promotive measure is to give financial and tax incentives for transfer of technology, regarding imported technology as a component of foreign investment to be attracted. Since biotechnology is a fast-developing field, it appears that, for developing countries with the goal of biotechnology development, promotive measures are, in general, to be preferred to control measures for regulation of technology transfer. Provisions should however be made of the unpackaging of technology in the long run.

As already stressed, effective transfer of technology followed by adaptation and development, requires continuous interaction between the technology receivers and R&D personnel. This interaction is lacking in many developing countries. The technology receivers are mostly in the private sector, often closely tied with transnational corporations through joint venture. The proprietary nature of the technology works against interaction with R&D personnel, by far concentrated in the universities and the public sector. The possible additional role of the universities as intermediary institutions for effective technology transfer has mostly been unexplored. Some developing countries are, however, looking into this role more deeply. On the one hand, placing industrial science and technology parks in close proximity with the universities, and on the other hand, forming units for commercial contracts in the universities. Some developing countries also have government-sponsored institutions for contractual R&D and other activities of benefit to technology transfer in the private sector.

It has been suggested (see e.g., Ventura, 1982) that developing countries deploy similar strategies as do some large companies in industrialized countries in acquiring biotechnological capability. These companies, lacking initial expertise and know-how, opted to buy out, or buy into, small companies which have been successful with biotechnologies. It was argued that by becoming integral parts of small, ambitious, energetic operations, developing countries would have direct access to expertise, information, training opportunities, and can also influence the R&D directions in these companies. This argument is attractive to a certain extent, but great caution and forward planning will be needed over selection of the areas and specific companies in which to make the investment, and the mechanisms for transfer of technological capabilities.

PATENTING SYSTEM AND PROTECTION OF INTELLECTUAL PROPERTIES

As the patenting system in industrially developed countries is rapidly undergoing adjustment with respect to protection of biotechnological inventions, a striking contrast is seen with the developing countries, only few of which have patent laws. Patent laws which exist in the few developing countries, not surprisingly, reflect the status of industrial development. The laws are therefore not specifically suitable for protection of biotechnological inventions, with which the developing countries are still unfamiliar. Furthermore, national policies in some countries dictate against patenting in certain areas, e.g., drugs, plant varieties, etc.

Although a main purpose of the patenting system is to further the early disclosure and wide dissemination of technical knowledge embodied in the invention, there is a reluctance of many developing countries to strengthen the system so as to enable a more effective protection of the inventors' intellectual property. This is due to fear that the enhanced protection will hamper endogenous technological and industrial development. This fear is reinforced by the fact that the great majority of patents in a typical developing country are granted to foreign inventors.

Biotechnological inventions present unique problems concerning protection for developed and developing countries alike. These problems arise mainly due to the questions over which products from and processes using living organisms, constitute patentable inventions, and whether the living organisms themselves are subject to patent protection (Beier *et al.*,1985; Straus, 1985). As commercial biotechnology continues to make an increasingly greater impact, these problems are being answered through landmark court decisions, and national and international measures. There are, however, still numerous legal differences among various industrially developed countries concerning protection of biotechnology. International harmonization on such protection would be a desirable stimulus for international development of biotechnology. The interests of the developing countries should also be taken into account in the attempt to achieve such harmonization. National upgrading of patent protection in biotechnology, as well as in other industrial fields in general, should serve the interests of the developing countries in the long run, since history has shown that an efficient patent system is favorable to technological and industrial development. A gradual approach to the upgrading of protection may be necessary, so as to achieve maximum benefits in line with the existing infrastructure.

UTILIZATION AND CONSERVATION OF ENDOGENOUS BIORESOURCES

Many developing countries, especially those in the tropical belt region, are rich in natural bioresources on which the biotechnological processes are based. An important policy target for these countries, therefore, should be the utilization and conversion of these resources through biotechnology. New developments in protection of intellectual properties in industrialized countries, including protection of micro-organisms and plants originally occurring in the developing countries, but having been isolated or modified by foreign inventors, pose an important dilemma for the developing countries over policy for most effective use of the bioresources. To open these resources to the international industrial community would, on the one hand, encourage more effective utilization through biotechnology. On the other hand, such an open policy might inadvertently lead to restriction of use by the developing countries of their own resources in the long run. Depending on the status of biotechnology and bioindustry, each developing country has to find the optimum measures which will allow both the growth of technology and the retention of ownership of its endogenous resources.

The use of a few superior resources, say, genetically improved plant species, has a side effect which must not be ignored by the developing countries. As the extent of this use grows, the survival of the various original genetic resources is threatened. Conservation of the original genetic resources should be an important aim of biotechnology as well as development and utilization of improved resources. The conservation is important not only for a passive purpose of maintaining the original gene pool, but also for the active purpose of a safeguard against inadvertent loss of genetic characters which might be needed in unforeseen stress situations or in further improvement of the resources.

SAFEGUARDS AGAINST BIOHAZARDS

Just as the questions of safety to the public, either from laboratory or field experiments or actual applications in biotechnology, are taken seriously in developed countries, so should they also be raised in the developing countries intending to promote their biotechnological capabilities. The developing countries are fortunate in that they can learn from the experience of developed countries. The safety of genetic engineering experiments and genetically engineered microorganisms, for example, has been subject to scientific tests and public debate for over a decade now, and the conclusions reached so far should be useful to the developing countries in devising safety guidelines. Ensurance of safe commercial products of genetic engineering is a part of necessary development work by the companies which must meet rigorous criteria before the products can be released. The issue of environmental impact of released biotechnology products is being actively examined. Developing countries should take an active interest in these issues, and take precautionary measures as necessary. These issues are important even for countries which do not have an active policy for biotechnology development, because they are potential importers of the products.

CONCLUSIONS

Various issues have to be raised for both national and international consideration in the effort to bring about biotechnology development for the benefit of the developing countries. Up to now, most of the benefit from new biotechnology is mainly concentrated in the developed countries, which have the advantage of existing scientific, industrial, and economic infrastructure. Yet, many developing countries have comparative advantage in containing rich bioresources as raw materials for potential products of biotechnology. In order to succeed in developing these bioresources specifically, or to gain a foothold in bioindustries in general, they need to have effective means of promoting buildup of required manpower and endogenous R&D capability as well as receiving transferred technology.

National efforts to achieve these goals may vary with the original status of biotechnology, and with the unique socioeconomic and industrial infrastructure of each developing country. Institutional measures which have been taken in selected examples examined here include establishment of national centers and network for support and implementation of important biotechnology projects. Policy issues concerning priority areas, linkage with the industries, protection of intellectual property, conservation of genetic resources and biosafety also need to be considered at the national level. Whatever mechanisms are taken, the participation of scientists, industrialists, and policy makers of varied backgrounds is needed because of the broad and interdisciplinary nature of biotechnology and its applications.

There have been some international efforts for biotechnology development for the benefit of developing countries. These efforts have sensitized a number of countries to the potentials of the technology, but so far have not resulted in substantive cooperative measures, although the attempts to establish the International Center for Genetic Engineering and Biotechnology deserve a mention in this regard. Future efforts are much needed both from the developing countries and the developed countries. For humanitarian reasons alone, the developed countries which have the biotechnological means to solve the problems of hunger, disease, and subsistence in general would be obligated to help the struggling developing countries. However, the input from developed countries toward development of biotechnology in the developing countries will not only benefit the latter exclusively, but will help the former in, e.g., utilization of resources, lowering production costs, and expanding markets. Hence, mutual benefit will be obtained for all countries in international cooperation in biotechnology.

REFERENCES

Beier, F.K., Crespi, R.S. and Straus, J. (1985). *Biotechnology and patent protection: an international review*. OECD, Paris, p. 133.

BOSTID (1982). *Priorities in Biotechnology Research for International Development*. National Academy Press, Washington, D.C., p. 261.

FAO (1981). *Agriculture: Toward 2000*. FAO, Rome, p. 134.

Ministry of Science, Technology and Energy (MOSTE), Thailand and BOSTID (1986) *Development of a National Center for Genetic Engineering and Biotechnology in Thailand*. Report of US Advisory Group Visits to Thailand, 1984. National Academy Press, Washington, D.C., p. 88.

Oei, Ban Liang (1986). *A Status Report on Biotechnology in Indonesia*. Abstracts of First ASEAN Science and Technology Week Conference, Kuala Lumpur, Apr. 1986. Ministry of Science, Technology and the Environment, Malaysia.

Straus, J. (1985). *Industrial Property Protection of Biotechnological Inventions*. World Intellectual Property Organization, Geneva, p. 89.

UN (1979). *The Long Step Forward*. The United Nations Conference on Science and Technology for Development. UN, New York, p. 21.

van Hemert, P.A., Lelieveld, H.L.M. and la Riviere, J.W.M. (eds.) (1983). *Biotechnology in Developing Countries*. Delft University Press, Delft, p. 158.

Ventura, A.K. (1982). Biotechnologies and their implications for third world development. *Technology in Society*, 4: 109-129.

Warren, K.S. (1985). Great neglected diseases of the developing world and their possible control by vaccination. In *Vaccines 85*. "Molecular and Chemical Basis of Resistance to Parasitic, Bacterial and Viral Diseases" (Lerner, R.A., Chanock, R.M. and Brown, F., eds.). Cold Spring Harbor Lab., p. 374.

Yuthavong, Y., Bhumiratana, A. and Suwanna-adth, M. (1984) The status and future of biotechnology in Thailand. In *Proceedings of ASEAN-EEC Seminar on Biotechnology: The Challenges Ahead*. Singapore, Nov. 1983. Science Council of Singapore, pp. 29-39.

Zimmerman, B.K. (1984a) The International Centre for Genetic Engineering and Biotechnology. *Bio/technology*, Jan. 1984, pp. 55-59.
Zimmerman, B.K. (1984b). *Biofuture, Confronting the Genetic Era*. Plenum Press, New York, pp. 133-139.

Biotechnology: The University Perspective

Robert Barker
Cornell University
Ithaca, New York

INTRODUCTION

The rush to commercialize products derived from genetically engineered organisms caught almost everyone by surprise, except perhaps the founders of Genentech. The various tools of what is now termed biotechnology were developed almost without exception by scientists working in universities, federal laboratories, and non-profit institutes. Although some technologies had been known for years, it was not until the late 1970s that it was generally known that a battery of these technologies could be brought together to create new life forms and cause bacteria (and presumably other organisms) to make new and potentially valuable products. Realization dawned simultaneously on faculty members in the biological sciences, on entrepreneurs and venture capitalists, and on the scientific managers of established industries. The prospect of a biotechnology industry, one with a predicted market impact of $100 billion by the year 2000, made for very interesting times in the resarch universities in 1981 and 1982.

Calmer times prevail now, but it is worth describing briefly some of the unusual opportunities that were opened to universities and to faculty members at that time. In considering the opportunities that have been opened to universities it is important to understand the institutional values and policies with which some of the proposed new ventures were in conflict.

THE RESEARCH UNIVERSITY: PRINCIPLES AND PARADOXES IN THE ADMINISTRATION OF RESEARCH

PUBLICATIONS AND THE PUBLIC NATURE OF RESEARCH FINDINGS

As non-profit institutions simultaneously engaged in research and education, most universities have adopted policies that guarantee prompt and full disclosure of research results through publication. In addition, new findings are often quickly transmitted among colleagues locally and more widely through presentations at seminars, symposia, and national meetings. Publications are vital. They are the principal means of evaluating a scientist's worth and are decisive in appointments, promotions, the awarding of tenure, and competitive grants.

Many universities forbid classified, secret, or proprietary research. Even those that do not, however, take great care to separate the normal research activities of the university from those activities that are restricted. In fact, it is almost impossible to mix graduate education and proprietary research. The reason is simple—graduate students must publish their theses.

PATENTS AND LICENSES

In what would seem to be a contradictory posture, most universities and most research sponsors encourage patenting and licensing of inventions that have commercial potential. Commonly, the university bears the cost of patenting and the inventor receives a percentage of royalties and/or income from licenses.

Universities differ substantially on the patent and licensing arrangements they will make with a research sponsor. Federal sponsors expect the university to patent useful inventions and to license them as appropriate (exclusive licenses are permitted). With private sponsors a university may arrange to transfer patent rights or to give exclusive licenses with or without royalties. Most research universities are anxious to ensure that inventions are patented and licensed and by that means enhance institutional income. No one objects to a faculty member or institution becoming rich.

The paradox of non-profit organizations pursuing inventions for profit is only made rational if it is understood that patents are not pursued as an end in themselves. They are seen as the inevitable and occasional consequences of basic research aimed at understanding the basic natural laws and their application in the sciences and engineering.

The philosophic conflict is greater in land-grant universities where the research mission extends to development, demonstration, and even delivery of a product or service; in most cases without the benefit of patent or license, although there have been some notable exceptions (warfarin and vitamin D).

DEVELOPMENT AND COMMERCIALIZATION

Realizing the economic potential of a basic discovery usually requires substantial investment for development. And this can occur only through the private sector. If university research is to stimulate economic development, it is essential that effective linkages be formed so that industry can draw on that research. In most cases, universities must avoid pursuing development, yet ensure that basic research, even when it does not result in patents, is available to industry for development toward commercializable products. Universities must walk a careful line between the loss of economic benefits to society and the inappropriate use of university resources in pursuit of a profitable outcome.

A second feature of university life that is important to understand and that also has elements of paradox is the provision of released time to faculty members for consulting. Most allow one day per week for consulting. Faculty members are encouraged to participate as members of corporate advisory boards and as expert consultants to industry and to federal and state agencies. As members of an open community in which colleagues share new findings readily, faculty members acting as consultants to industry must be careful not to make inappropriate use of information that may be viewed as privileged by those who provided it. Interestingly, not all faculty members agree information shared before publication may be privileged. To some it is public property. In areas, such as biotechnology, where basic discoveries may translate quite directly into application, the hot new finding shared over a cup of coffee with a member of the community who is also a member of a corporate advisory board, or a consultant, may result in the economic benefit of that discovery being realized by non-contributing individuals and organizations.

INDUSTRY: FROM VENTURE CAPITALISTS TO CORPORATE GIANTS

VENTURE CAPITALISTS: LIMITED PARTNERSHIPS

In 1981, at Cornell University, venture capitalists came in $30 million packages. At least most came suggesting that, if the University would lend its name and help organize its faculty, it would be possible (even easy) to raise $30 million to create a limited partnership to fund university-based research. The payoff to investors would be patents and the royalties therefrom.

Given the historically poor performance of most patents (very few have generated more than $1 million in their lifetime), it seemed likely that most investors in these limited partnerships would reap no more than a tax write-off. To be sure the venture capitalists, in the tradition of the trade, would be involved conservatively and would benefit in the transaction. The university would benefit in the flow of research funding. In many cases there was a clear expectation that university research would be targeted to enhance the likelihood that patents of value would be generated.

VENTURE CAPITALISTS: NEW CORPORATIONS

A different approach, and a much more successful one, was taken by venture capitalists who undertook to create a new corporation starting with a scientific advisory board of leading scientists (mostly from universities). This mode led to the development of Biogen, Plant Genetics Inc. and perhaps other corporations. Starting with a relatively small investment (say, $1 million), the advisory board was formed. Its members each were given a small percentage of the new corporation (say two percent) and charged to produce the scientific business plan for the corporation. The targets for commercial development were identified together with the laboratories best situated to pursue the project. At this stage, new funding was sought and the research development effort began through contract research. Later, when these development efforts indicated a likely prospect for commercialization, the second private offering of stock was made, providing sufficient funds to begin the commercial venture. Public offerings were made later. The importance of this strategy to this discussion is the new role in it for university scientists and the very tight linkage between basic research and commercial outcomes. These arrangements, and others like them, resulted in faculty members having very substantial financial interests in the success of the corporations they advised. A successful venture could mean several million dollars to each adviser. The value of new knowledge critical to the commercial enterprise was great. Colleagues and students were sources of information, and research strategies and targets, supported by public funds, could become a means to try to ensure a successful venture. In some cases, the conflicts of interest and commitment for these faculty members became so intense that ties with the university or with the corporation were severed.

THE ESTABLISHED CORPORATION

With few exceptions, during the 1960s and 70s research intensive corporations had paid little attention to areas now known as biotechnology. It was clear by 1980, however, that the pharmaceutical, medical diagnostic, agricultural, energy, and chemical industries would be greatly affected by the new technologies. Most corporate managers knew little of the field and there was a rapid escalation of interest in university-based research. Several major corporations moved quickly to form collaborations with universities. Monsanto with Washington University, Hoechst with Harvard, DuPont with MIT, all looking for a window on biotechnology; but each arrangement differing in the extent to which industry and university are mingled in the research effort.

Most research universities were as anxious as the research corporations to develop collaborative programs. It seemed to university scientists and administrators that most major corporations were willing (even eager) to make multi-year, multi-million dollar commitments. There was a genuine interest among the scientists in establishing programs that would help ensure that their work was used to good purpose. And for some, joining with a group of colleagues in a collaboration with a major corporation (or preferably several simultaneously) appeared to eliminate some of the troubling issues raised by other and more venturesome approaches.

In what follows, I will deal with some features of university-industry collaborations that are essential to their success.

UNIVERSITY-INDUSTRY COLLABORATIONS IN BIOTECHNOLOGY

BASIC PRINCIPLES

I take as given that:

- The basic research in biotechnology carried out by university scientists has been well and effectively done and that university-industry collaboration should not redirect this research toward development and application.
- Universities should remain "open," research and scholarship should lie in the public domain, information should flow freely, and there should be easy access to the campus.

- Individual faculty members should be free of substantial conflicts of interest and/or commitment.
- Research, especially that carried out by undergraduate and graduate students, should be published.
- Substantial federal funding of investigator-initiated research will continue to provide most of the support for basic research related to biotechnology.

FINANCES

Federal Funding—Traditionally, funding for basic "biotechnology" research has been provided by federal agencies. This continues to be the case despite the influx of industry funding in the past few years. With federal funding at more than $500 million per year the proportion provided by industry may be as low as one in 25. To the extent that industry funds target the same or related opportunities as do federal funds, there is substantial leverage.

As the need for close collaboration between universities and industries has become more apparent, state and federal agencies have begun to fund programs that require industry participation. Leverage, and being leveraged, is becoming a way of life for university scientists. But even these additional funds do little to change the proportion of funding that supports basic research.

Corporate Funding—Why then do universities compete so vigorously for corporate funding? In most cases the funds are valued for the prestige they bring, because they form a linkage that the university values, because they are seen as addressing a national need and, most important, they allow new dimensions to be added to the basic program. In many cases industry funds are used to bridge between disciplines and to bring biotechnology tools to bear in areas that are not of interest to federal agencies. Finally, as pointed out above, some federal and state funding is tied to industry participation.

The provision of some funds by industry probably is essential to a successful program of university-industry collaboration, even when state or federal funds are available for similar purposes. Such funding:

- Assures industry commitment.
- Provides added resources that allow universities to commit faculty and staff time to the program.
- Allows new fields of interest to industry to be explored.

- Provides an equitable basis for partnerships.

Industry funding, as with all sponsored research funding, can only be accepted on the basis of the established principles of the university. At Cornell this means that no proprietary research will be done, publication decisions lie with the responsible faculty member, patents will be held by the university, and the university decides who may work on the project. For many collaborations the key issue is the ownership of patents and the control of publications, matters of intellectual property rights.

New Venture Funding—Many of the new corporations established in the biotechnology industry are small. They are research and development intensive and could use the assistance that could be provided by university scientists. To my knowledge, the only new ventures that have been able to collaborate with universities have been of the kinds described earlier under the section, VENTURE CAPITALISTS. Those new corporations that are not beneficiaries of a high initial level of capitalization have no means of purchasing the assistance they need. Neither have universities developed effective methods of working with small corporations in what would have to be a venture mode for the university.

TECHNOLOGY TRANSFER AND INTELLECTUAL PROPERTY RIGHTS

Technology Transfer—University-industry collaborations are undertaken to enhance and ensure "technology transfer" from the university to industry, although in many cases transfer in the reverse direction doubtlessly occurs. Technology transfer here refers to all forms of information transfer. The bottom line is to increase opportunities for economic development, whether that be through education or the exclusive licensing of a new method for gene splicing.

Technology transfer includes a full range of communications and only works when both the university and the industry commit the human effort needed to ensure success. It is a people-intensive activity. When a single investigator is involved in a well organized research project communication with the industry counterpart is direct and effective. When a broad-based program is involved, communication is more difficult. It cannot be left to the university alone, industry must bear an equal responsibility. A successful program will place heavy responsibility on the program director and will involve formal and informal contact, extended visits by industry scientists to the campus, frequent visits of university scientists to industry laboratories, and various written forms of communications. In the fully successful program, drafts of publications will contain no information that is wholly new to the industry participants.

Intellectual Property Rights: Patents and Licenses—A fully effective communication program may seem to sweep away concerns with intellectual property rights. After all, if everyone knows what is going on, what concern can individuals or the participating organizations have in such rights? But the problem remains as perhaps the major one to be resolved in university-industry collaboration. Who shall hold the patents? Who can or cannot obtain a license?

Most universities have patent policies that require faculty members to give the university first refusal of inventions that might be patentable. Those the university refuses can be pursued for patenting by the faculty member. In the former case, royalties accrue to the inventor(s) and to the institution, often split between the department, college, and university. Faculty members have a vested interest in successful inventions and so does the university.

The patent problem is dealt with most simply when a single investigator works with a corporation, and it can be demonstrated that no federal funding is involved in the project. Here the university can cede the patent rights to the corporation, grant a royalty-free exclusive license, or make whatever arrangement fits the case. When federal funds are involved, the university must hold patents but can arrange to license as it deems appropriate. In both cases with the concurrence of the investigator.

The multi-corporation, multi-disciplinary program with many individual projects is more difficult. Because of the high level of federal funding, patents must be held by the university. Indeed federally funded research may have been the basis for the university's attractiveness to the corporations. How to license patents poses a dilemma. Not all inventions will be equally interesting to each participating corporation. Depending on the significance of the invention, it may be best for each participant to have a royalty-free, non-exclusive license. In other cases, royalty-bearing exclusive licensing to a single corporation may be best. I know only the Cornell program well. In it decisions on how to deal with licensing were left to a case-by-case review and decision by an executive board whose members were drawn from the university and the participating corporations. The university and faculty members chose to take a relaxed view of their own rights for royalties.

Publications—Publication of research findings is essential for university research. Decisions on when and what to publish must lie with faculty members. In most cases universities can accept programmed delays in publications of 30, 60 or 90 days (journal review procedures take at least that long) but must leave decisions about longer delays to the scientists. I know of no case in which a faculty member has insisted on publishing when that would compromise obtaining a patent.

Industry Participation in Collaborative Research—Again, the single project is easily arranged. It can involve university and industry scientists in whatever fashion suits them within the guidelines set by the university and the corporation. The work can be done at either site by the participants best qualified to do it. In most cases, however, the work is done at the university by faculty members and students. Only rarely does the industry scientist participate directly, acting instead like a grants officer with a higher than usual personal interest in the project.

Broader, multi-project programs require a continuing dialogue between the corporation(s) and the university. Ideally projects should reflect the shared interest of contributing partners (including the university) and funding should not go to projects that are competitive for funding by federal agencies. This is not an argument for funding less deserving projects but to fund new, often multi-disciplinary projects. These are often more difficult to obtain federal funding for, since agencies have narrower missions and tend to be highly conservative.

At Cornell, where three corporations (Kodak, General Foods, Union Carbide) and the state of New York provide a total of approximately $2.5 million per year, a request for proposals is issued twice a year. The criteria are the fundamental nature, novelty and multi-disciplinary character of projects. All proposals are reviewed by a Scientific Advisory Committee drawn from the university and the participating corporations. These proposals have proven to be a most effective way of communicating to the corporation where the cutting edge of biotechnology is in the view of a substantial cross-section of the faculty. Approximately 20 percent of submitted proposals have been funded, most have involved more than one faculty member. Effects of the program on collaboration in research and teaching have been substantial.

An important feature of the Cornell program is the participation of corporate scientists as resident scientists on campus in basic research in collaboration with one or more faculty members. The first resident scientist stayed two years, others are being sent for a year. These resident scientists play an essential role in communication. During their stay on campus, they return regularly to their corporations and serve as consultants in ways that few hired consultants can. Though they do no proprietary research, they provide the best conduit for technology transfer. In all cases the resident scientists have developed new skills that allow them to enhance the biotechnology programs of their parent corporation.

BIOTECHNOLOGY: IMPORTANCE TO THE RESEARCH UNIVERSITY

The technologies that are now captured under the rubric "biotechnology" grew out of fifty years of basic biological research carried out in large part by university scientists. They did not set out to create these technologies with a view to their potential for economic development. Instead they were focused on questions of how cells work and how genetic information is used. Analytical tools of great sensitivity and precision had to be developed along with strategies for dealing with single cells and complex biomolecules. The importance of "biotechnology" to universities lies principally in the value of these tools in the continuing quest to understand living systems. The spread of biotechnology into every field of biology has been extremely rapid and its application holds even greater promise there than in the marketplace.

But there are other ways in which biotechnology will be important to the research university; among them:

- As a basis for continuing interaction with industry.
- As a new applied science, in essence as biological engineering.
- As a catalyst for new development in agriculture, medicine, food science, and engineering.

All are related, all bring with them challenges to the universities and to the agencies that sponsor basic research in the biological sciences, including:

- Finding the means to build strong, long-term relationships with industry that respect the principles of the open university.
- Providing the facilities needed to allow the application of biotechnology approaches in many new areas of the biological sciences.
- Maintaining strong research and graduate education programs in the areas fundamental to the future of biotechnology. These include cell and molecular biology, biophysics, biochemistry, and genetics. Faculties, facilities, and finances will all present problems.
- Maintaining a balance between basic and applied research.

The industry connection is worth further comment. It is important for several reasons. As a new field, biotechnology offers the opportunity for universities and industries to work together as the industry grows, forging a joint effort that partitions basic research and application on a partnership basis. Economic development is, in part, a responsibility of the universities. They must collaborate with industry and, perhaps especially in biotechnology where basic research can translate quickly into application, should protect their role in basic research programs by establishing effective mechanisms for continuing technology transfer. In these relationships, industry should see the university-based program as the conduit to the total spectrum of basic research and, at the same time, as the beginning of a continuum leading through corporate development to commercialization.

BIOTECHNOLOGY: IMPORTANCE OF THE RESEARCH UNIVERSITY TO INDUSTRY

Current federal annual expenditures for basic research in biology may total close to $1 billion. I exclude here much of NIH funding for biomedical research which, if it was included, would bring the total close to $4 billion. Of these amounts, more than $500 million supports work directly relevant to biotechnology. The rate of scientific progress is rapid. Discoveries made using one organism are transferred quickly to others. The field cannot be partitioned easily, requiring those concerned with commercialization to stay current with the broad field. No corporation can afford to duplicate enough of the federal effort in-house to be able to ignore university collaborations. All must have effective interaction with university scientists. They will differ only in how they choose to do it.

Those who make best use of the lesson of the last decade, that basic science can generate very substantial and unexpected economic outcomes (biotechnology), will find less traditional ways and form close working relationships. The approaches used by Monsanto with Washington University and Cornell with its collaborators serve as models.

These strategies have many positive features for the corporations but they can succeed in the longer term only when a corporation is able to adjust its social structure to accommodate the university interface. Among the positive features:

- Access to knowledge in a timely way.
- Enhanced technology transfer.

- Opportunity to help set university research priorities.
- Access to young scientists in training.
- Collegial relationships between corporate scientists and a broad cross-section of the academic scientific community.
- Continuing education of corporate scientists is built into the program.
- The university is a vehicle for communication with other public sector research institutions (universities, federal laboratories, private institutes).
- Access to university-based equipment and facilities some of which may be too expensive for a single corporation to afford.
- Provides a neutral ground for collaboration with other corporations. Such collaboration with other corporations and federal and state agencies can gain substantial leverage on expenditures for basic research.

All of these positive features can be obtained only if corporations accept that universities are public institutions, that their research and scholarship should be available to all, and that the university research mission should be for the long-term. In university-industry collaborations, corporations must accept the paradox that openness can be maintained while pursuing research that supports economic development.

Ten years ago biotechnology took the corporate world by surprise. This happened because the cultures of universities and industry had become separate in part because of their different ways of treating knowledge. Too tight a requirement for sequestering new knowledge could lead again in ten years to a separateness that will be to the disadvantage of both.

PART V
COMMERCIALIZATION ISSUES

Products and Health Care Delivery

Gary R. Hooper
Genentech, Inc.
South San Francisco, California

It's a real pleasure to be here today for the symposium, "Biotechnology: Perspectives, Policies and Issues." Since this section addresses commercialization issues of biotech products, I thought I would share with you some of the issues we have faced over the years, beginning with the selection of recombinant DNA products, move into the development and manufacturing of these products, and conclude with regulatory and clinical issues which must be considered prior to successful marketing.

In the time allowed, it will be impossible for me to cover all the numerous issues and problems faced by companies, like Genentech, involved in biotechnology. Thus, I will focus on a few areas which most frequently arise in conversations with people looking at the commercialization of recombinant DNA technology.

However, in order for everyone to appreciate the decisions we have made in commercializing our technology, I must say a few words about the technology and the goals of Genentech. Obviously, decisions which we have made over the years are tied directly to who we are, where we are, and where we want to be in the future.

Who is Genentech? In very simple terms, Genentech is a company which looks at a human disease and determines what is naturally found inside the human body which will either treat or cure the disease. Next, we isolate the substance, determine its structure and locate the gene which codes for the production of that human protein. As you may recall from your high school science class, a gene is a blueprint, or set of instructions, that tells a cell what protein to produce. Once we locate the gene, we remove it from the human cell and place it inside microorganisms such as *Escherichia coli*, a bacterium found in human intestine, or saccharomyces, simple baker's yeast.

Now we have microorganisms that contain the blueprint for the protein we desire. These organisms are then grown through a process of fermentation. In fermentation, we provide an environment that is conducive for the organisms to grow. As the microorganisms proliferate and use their own genes to produce proteins required to sustain their lives, they also manufacture the protein we desire by using the gene which we have placed inside. We then kill the microorganisms and isolate the desired protein from all other proteins. Thus, Genentech is in the business of creating little chemical factories. Chemical factories that produce therapeutic proteins.

It is important to realize, in spite of all the excitement and claims made in recent years, that this technology is just another manufacturing process. Products coming out of the technology must compete with other products and forms of therapy. Thus, marketing, distribution, etc., will still play a vital role in the success of any biotech company.

In the above description, I focused on human therapeutic proteins. Genentech's goal is to conduct research, manufacture, and market therapeutic human products. We are not a diagnostic company. We are not an instrument company and we are not an industrial chemical company. True, we have joint ventures in these areas, but they were established to capture value in spinoff products generated while pursuing human therapeutic proteins.

With all of the potential and numerous applications of recombinant DNA technology, why focus on human therapeutics?

- What better way to demonstrate the benefits to be derived from this new technology than to treat diseases that were previously untreatable or to help someone with a serious health problem.
- Margins are high in therapeutic medicine relative to most other industries. This would allow us to fund our own growth and be the established pharmaceutical company that has always been our goal.
- Hospitals are definable and accessible. There is a significant amount of information on U.S. hospitals which tells us where they are located, the types of illnesses they treat, patient populations, etc. This reduces marketing costs which is very important to a new company.
- It is relatively easy to identify target markets within the hospitals. Again, this reduces marketing costs.
- You do not have to create a sales force to call on all U.S. hospitals. By limiting your target audience to critical care facilities you are able to obtain good coverage with 50 to 80 sales representatives, instead of the 600 to 800 that would be otherwise required. It is significantly more economical to support 80 sales representatives than 800.
- Therapeutic proteins offer good expansion opportunities. Some therapeutic proteins are known, but this is just the tip of the iceberg. There are many proteins whose functions are unknown and many proteins not yet discovered. Some of these will have high therapeutic value which can be captured by this technology.
- By focusing on critical care facilities you have fewer drug delivery

problems. Therapeutic proteins must be given by injection. Patients in this type of facility are used to injectable medicine, versus taking a pill from your local pharmacy.

- Finally, the financing of a young company through its early years is a high-risk investment. Large pharmaceutical companies are familiar with high-risk investments in research and development. Thus, contract research with large pharmaceutical companies would enable us to maintain our growth until survival was assured through our own products.

From the start, Genentech's desire has been to manufacture and market its own therapeutic products. With this in mind, we can now move ahead with a historical view of how we handled the challenges of commercializing our biotech products.

On numerous occasions I have been asked how we strategically select our therapeutic products. There is no mysticism associated with product selection. In fact, it has been primarily dictated by the needs of the company at that particular stage of growth, yet keeping in mind our corporate goals.

For example, when Genentech was formed, its first task was to prove that a human protein could be produced in a microorganism. At that time, most people did not believe this was possible. After careful consideration, somatostatin was selected as our first project. This is a hormone, produced in the brain, that plays a regulatory role for numerous other hormones and was believed to be useful in treating certain disorders caused by a hormone imbalance. Somatostatin was chosen because it is a small protein whose structure was known and had potential therapeutic value.

Needless to say, the project was successful and demonstrated the technical feasibility of the technology, but it did not demonstrate financial feasibility. Some people also questioned the usefulness of the technology. The next project selected had to not only demonstrate the financial feasibility, but also illustrate the good that this technology represented to mankind.

The project selected was human insulin, which again was successful and is currently being marketed by Eli Lilly. This set the stage for our continued growth as a company with greater flexibility in product selection. We next selected projects of known therapeutic value, with worldwide markets already in existence. These were selected for several reasons. Market introduction would be less expensive since distribution channels were already in place, location of patients was shown, a shortage of product existed so demand was high, correct dosage was already established, and we may only have to show bioequivalency in clinical trials. In addition, we could obtain additional revenue through the licensing of foreign partners. We believed it would be easier to license known products with known markets than totally new products with unknown potential. Human growth hormone is an example of this logic.

With this philosophy of product selection, Genentech was able to build a solid foundation of therapeutic projects while significantly advancing the science. Soon we were in a position to expand our product portfolio to include high-risk projects, those with a higher risk of scientific success and therapeutic value. Our product selection criteria were again modified. However, relative to large pharmaceutical companies, these were very simple.

- There must be a significant critical care need. Either current therapy is inadequate or existing products were in need of improvement.
- The project had to be scientifically feasible and be completed in less than two years, that is, from beginning the project to expression of the protein in a microorganism. Today, this often takes only a few months.
- The product must be highly potent in small quantities. This means that a smaller manufacturing plant would be required which translates into less of an investment. Capital expenditures are very important to a start-up company like Genentech.
- The market potential had to be greater than $100 million. If the project met the other criteria, we were not concerned about how much over $100 million it may be worth. It was sufficient to be of interest and the marketing group could determine the actual potential at a later date.
- Finally, it had to be scientifically romantic or sexy. In other words, it had to be extremely interesting and challenging scientifically. We had to further extend our scientific lead and prove that difficult proteins could be produced through recombinant DNA technology.

We therefore chose projects such as alpha, beta, and gamma interferon, and factor VIII. These projects increased our technological learning, validated the quality of Genentech's science, and attracted additional funds for our continued growth.

Today, we are considered large, relative to other biotechnology companies, but we still view ourselves as small. As I mentioned earlier, our technology is just another manufacturing process. Thus, in terms of other pharmaceutical companies, we're quite small. However, it is interesting to note that as we have grown to face more complex issues as a company, so have our project selection criteria become more complex. As we have more and more direct contact with regulatory bodies involved with biotechnology and as we face more and more competition, some of these lessons have filtered into our selection criteria. We now use terminology such as unique therapeutic niche, lead time, patents, acute or chronic diseases, immunogenicity, and toxicology. But our selection process remains simple.

Some time ago I had the pleasure of meeting the founder of Benihana Restaurants. On that occasion I asked him how he determined where to locate his restaurants. He said, "I follow the big boys. I let companies like McDonalds, successful hotel chains, and shopping center companies spend millions of dollars on research to tell them where to locate their outlets. Then I locate my restaurant down the street or around the corner."

Some pharmaceutical companies have complex formulas to select projects. Some have page after page of selection criteria. But it all boils down to a few key facts. In one sense, we are like Benihana, we follow the big boys. But we don't use complex formulas or a small book of criteria. We focus on those few key facts that are the real basis of decision making. There are three main considerations we use in looking at new projects with two or three questions in each area. Our main considerations are: (1) Scientifically is it feasible? (2) Medically is this the protein of choice? (3) From a business point of view does it make sense given our stage of growth and where we want to go as a company? That's it!!! No magic, just logical thinking!

With this simple outlook we selected projects such as Tissue-Plasminogen Activator, Tumor Necrosis Factor, Lymphotoxin, a number of growth factors, receptors, and immunomodulators. We are also investigating other proteins whose functions were previously unknown, but they all show potential value as critical care therapeutic drugs.

Thus, our selection process has evolved in parallel with the needs and goals of the company. We moved from proving the science and financial feasibility to more challenging projects which extended our scientific knowledge. We then looked at new and unknown proteins and now have expanded this horizon to redesigning old proteins to improve existing properties or add new and beneficial characteristics.

In reviewing the evolution of our product selection process it is obvious that we do not believe any selection criteria should be cast in concrete. On the contrary, it should remain flexible and work in concert with the needs and goals of a company. It must complement each stage of a company's growth and the company must be willing to adopt new criteria when required. This has been very important to us in the past, and I'm sure it will continue to be so in the future.

Once you have successfully completed your initial research and now have a microorganism producing the desireable protein, there are new issues which must be faced—mostly scientific. For example, the microorganism you now have producing the protein may not be the optimum host/vector system. Issues such as bioactivity, stability, glycosylation requirements, yield, etc., may cause you to shift from *E. coli* to another bacterium—or to yeast—or mammalian cell culture. Selection of an optimum host/vector system can become very complex. As a general rule, we will place the gene into all of our host/vector systems and select the one which is optimal. Selection considerations are bioactivity, yield, purification problems, stability, etc.

Once you have selected the optimum host/vector system you are now looking at scale-up and manufacturing. Here we have a whole new set of issues. Depending upon the host/vector system, your manufacturing process may be microbial fermentation or continuous mammalian cell culture.

With microbial fermentation, our first problem was to determine whether known technology was sufficient for fermenting our microorganisms. Let's face it! In the past, we had always been trying to kill bacteria, not grow them in large batches! We soon discovered that new technology had to be developed. Once this technology was developed, we faced problems in purification. New assays and purification systems had to be created. There was a definite shortage of assays that could be used for locating human proteins in a soup of bacterial proteins. Likewise, there were no isolation methodologies in existence for purifying human proteins from bacterial proteins. Obviously, you could not have a human injectable which contained bacterial endotoxins or bacterial proteins.

In continuous mammalian cell culture the issues are different. Here you are using an immortal cell line that does not die out after 50 generations or so as with normal human cells. Thus, you must be able to demonstrate that the final product contains no substances like viral RNA or DNA which could make a human cell become immortal and thereby cause cancer. Mammalian cell culture is also a slow and expensive process. Only high-value products can be produced using this process. For example, a product such as albumin that costs a few dollars a gram would not be economical in a tissue culture system.

There are also some situations where your selected host/vector system cannot be grown in large scale. The yield per cell may actually decrease when grown in large volume. Thus, you may have to change your vector system and start again.

Scaling up your manufacturing system is a long and laborious process and is actually a new research project for each product. For example, the bacterium-producing alpha interferon, when used to produce gamma interferon, requires an entirely different fermentation process for large-scale manufacturing. So for each product you must optimize the manufacturing process.

This also causes some management problems for a biotechnology company. Since each manufacturing process is a new research project, the primary people involved are research scientists. However, the basic nature of pharmaceutical manufacturing is to do the same thing over and over again as exact as possible. At some point you must freeze your manufacturing methodology and say, "This is what I am going to use as my commercial process." Meanwhile, the interest of the research scientists developing the process is to look for continued improvement. They will always want to make it better. This is a serious issue because you want to keep morale high and not inhibit the creativity of the scientists.

This leads me to another serious issue which we have had to face. How do you deal with the fast technological changes we continue to experience in recombinant DNA technology? Not long ago I purchased a home computer, but by the time I learned how to use the thing, better and less expensive computers were being introduced. This also happens in all areas of our technology. How do you take advantage of technological advances while maintaining a sense of continuity? Do you go with an intracellular or secretion process? Should you incorporate the manufacturing improvement now or wait until later? Even information obtained in clinical trials can have an effect on prior work.

Change also comes from outside influences. Changes in the Recombinant DNA Committee guidelines; changes in Food and Drug Administration (FDA) guidelines; changes in FDA reviewers and advisory committees all have an effect upon your operation and you must deal with this change. It is not an easy task. Meanwhile, your projects may be delayed from your estimated introduction time frames. You can no longer count on those revenues coming in as planned, but you still must survive as a company.

Even though these are serious issues we currently have and continue to face, we must all realize that it is a learning process for the various regulatory agencies as well as ourselves and those agencies have been very good at working with us to ensure a safe and efficacious product. It takes time, but we will all be better off in the long-run and that is what it is really all about.

This brings me to another very important subject. What are the regulatory implications of commercializing a recombinant DNA injectable? I have already mentioned some of the concerns like changes in regulatory people and guidelines, but there are others. For instance, how do you perform preclinical and clinical testing of a recombinant DNA product?

We are dealing with proteins. And some human proteins are species specific. In other words, it will not work in most other animals. This makes it very difficult to conduct preclinical animal tests. Some proteins will not work in standard mice, rat or guinea pig systems and you must go directly into primates. Testing becomes difficult and very expensive. Sometimes you must also produce the equivalent animal protein, such as rat gamma interferon, to test prior to moving into higher animals. Again, this is time-consuming and adds cost.

When you finally move into clinicals your problems again change. We are all interested in safe and efficacious product. But, who sets the gold standards? Is it the advisory committees, divisions of the FDA or individual regulators? Which office of the FDA? Take Tissue-Plasminogen Activator for example. Is it a drug, a biological, or a blood product? Arguments could be made for each.

In addition, what endpoint do you use to establish efficacy. Do you use drugs now on the market as a standard, or test against a placebo? Is it a biochemical or physiological endpoint—or do you use mortality, morbidity? For a recombinant vaccine, do you use sero conversion or disease reduction? Many of these issues are not unlike those faced by traditional pharmaceutical companies. However, they become more complex when dealing with a recombinant biological. As with most other drugs, there are really no set rules to be followed. Each product has issues which are specific to the protein and must be handled on a case-by-case basis.

I have tried to focus on some of the major issues and problems we have faced as a genetic engineering company by following the progression of products from research to marketing. I began by sharing with you our product selection process—then identified some of the issues we have confronted in moving products through the pipeline to marketing. Of course, it is impossible for me to address every issue, but I hope that this representative sample has left you with an appreciation of the complexity of commercializing a recombinant DNA product.

Changing Role and Scope of Venture Capital in the Commercialization Process of Biotechnology

Scott A. Bailey
Southeast Bank, N.A.
Miami, Florida

INTRODUCTION

"A well-developed life science base, the availability of financing for high-risk ventures and an entrepreneurial spirit have led the United States to the forefront in the commercialization of biotechnology." (U.S. Congress, 1984).

This congressional document continues with a discussion of the factors potentially important to international competitiveness in biotechnology. Of the three most important factors, two are discussed in this paper. One is the availability of venture capital to start new firms. Another important factor, personnel availability and training, is also discussed. This is done from a business and management perspective, rather than for technical assessment. One of three factors of moderate importance in the Office of Technology Assessment document is university/industry relationships. This concept, too, is discussed.

The author used the review of literature technique in the writing of this paper. Several points of view are presented to support notions of the author (Bailey, 1978) in his unpublished manuscript (which was abstracted for publication in *The Journal of Commercial Bank Lending* in 1979). In those writings, he described the techniques employed in financing the computer peripheral industry in the early 1970s. In many ways, that industry then closely resembled the biotechnology industry of the 1980s. This paper goes on to support the view that a history of financings and entrepreneurial activities was really a prologue to the future.

TRADITIONAL VIEW OF VENTURE CAPITAL

VENTURE CAPITAL

Venture Capital is defined as the financing of relatively small companies before they could qualify for public underwriting (Rubel, 1973). Practically all the biotechnology companies in existence had been the recipients of venture capital. The money invested is certainly necessary, but valuable management assistance from the financiers, as well as their contributions of effort as members of the Board of Directors, helped shape the corporation for its future growth.

Venture capitalists can be broken down into six categories, which are described below (Rubel, 1973):

Private venture capital firms. These firms consist primarily of professional partnerships and corporations that have been backed by institutional investors, such as insurance companies, pension funds, profit-sharing plans, bank trust departments, wealthy families, endowments, foundations and others.

Family venture capital firms. Although declining in terms of relative importance to the venture capital industry, family venture capital firms represent many of the nation's wealthiest families, and still play a significant role in the venture capital field.

Small business investment companies (SBICs). There are probably about 300 active SBICs at the present time (350 in 1986), and this group is growing in both size and activity. About half are oriented primarily toward venture investing, and the rest are interested in secured lending and real estate transactions.

Venture capital divisions of large industrial corporations. This is probably the latest type of venture capital organization to be formed, and has certain characteristics that make it different, yet appealing to biotechnology companies. These groups typically invest in situations when the product, market, or technology is related or of interest as a diversification opportunity.

Investment Bankers. A number of investment bankers maintain active venture capital operations, both for purposes of investing their own and client capital, and for engaging in corporate finance activity, involving the placement of private projects on a fee basis.

Other forms of venture capital. The variety of venture capital organizations also includes bank trust departments, insurance companies, pension and profit-sharing funds, hedge funds, investment advisors, individual investors, and venture capital subsidiaries of large financial institutions.

Each venture capitalist has different goals and philosophies about investing and returns required from portfolio investments. The principal objective is to maximize total return on the capital invested. In seed or start-up financings today, venture capitalists are looking for a 15 to 18 month period, at the most, from idea to initial product shipments. Family partnerships tend to participate in ideas that will benefit both the economy and society, while the venture capital arms of large corporations try to create merger and acquisition candidates for their parent.

Regardless of motivation, most want to see a profit of three to five times the investment (pre-tax) in a three to five year period. After all the winners and losers are tallied, the venture capitalist hopes to have a minimum compounded rate of return (after tax) of 15 percent on his capital (Rubel, 1973). In the environment of the 1980s, this minimum figure has increased to 20 to 25 percent return on capital.

STAGES OF CAPITAL

As the portfolio company moves through life from idea towards maturity, the different stages it passes through in its financing history are fairly common to all young companies. The venture capital industry has evolved a series of definitions which are generally used by everyone. One venture capitalist's definitions are listed below as an example (Fitzsimons, 1984).

Seed. Normally the earliest and the smallest amount of capital raised by a company. It is often provided by the entrepreneurs to fund development of the business plan, market research, and engineering designs of prototypes. The financing is rarely enough to recruit a full management team or make capital equipment and facility commitments.

Start-up. The earliest of the large scale fundings provided by outside sources, principally venture capital firms. This accelerates product development, management recruitment, and capital equipment process. It may be provided in multiple rounds as benchmarks are met. It is usually sufficient to make inventory purchases and fund the earliest product deliveries.

Early Stage. This round provides expansion capital to fund traditional working capital needs and losses. Shipments, marketing, manufacturing, engineering, and staffing functions are entering rapid growth mode. Cash flow is negative and supplemental bank credit may not be available. Initial market acceptance has been verified. The longer term outlook for the company and its product remains uncertain.

Growth. All corporate key ingredients are in place. Cash flow is negative to break-even. Bank financing is in place to sustain current operations.

Planned growth rates in shipments, sales organization, manufacturing, and staffing require supplemental capital. Longer term outlook improving. May be final private financing.

VENTURE CAPITAL IN THE BIOTECHNOLOGY INDUSTRY

Two distinct types of firms are pursuing the commercial applications of genetic engineering in the United States: the small start-up companies founded primarily since 1976 to capitalize specifically on genetic engineering research, and the established multiproduct companies in such sectors as pharamaceuticals, chemicals, agriculture, energy, and food processing that have invested in the field. The interplay between the two and the complementary efforts of each have done much to give the United States its current lead in biotechnology (Schoemaker, 1986). This paper will examine various forms of financing employed by companies in the biotechnology field.

FUNDS FOR BIOTECHNOLOGY

Between these two types of firms, considerable amounts of money have been invested in biotechnology. Since 1976, several billion dollars have been funneled into the start-up biotechnology firms, of which there are now more than 200 (Schoemaker, 1986).

Funds raised by U.S. biotechnology firms in 1985 exceeded $200 million. Research and development limited partnerships (RDLPs) accounted for 28.9 percent of this financing. Corporate investment accounted for 35.4 percent, and equity financings in the public markets made up the remaining 35.7 percent. Historically, funds raised in the early 1980s were in amounts less than last year's total. In 1983, the amount skyrocketed to almost $600 million. As the initial euphoria of the industry began to wane, 1984's funding amounted to less than $200 million. Conversely, with today's increasing investor interest, financings already exceed $300 million as of April 11, 1986. In total, American investors have staked more than $3 billion on biotechnology in the past 10 years (*Economist*, 1986).

According to Murray (1986), the total private investment in U.S.-based biotechnology through the end of 1985 was $4.004 billion. There have been primarily four distinct types of private transactions comprising financial commitments to biotechnology. Equity purchase was $2.581 billion

(65 percent), contract research and joint venture were $578 million (15 percent), RDLPs accounted for $558 million (14 percent), and grants to universities or research institutes made up $260 million (6 percent). Product license agreements totalled only $14 million, or less than 1 percent.

CORPORATE AND INSTITUTIONAL VENTURE CAPITAL

To finance their research and development efforts, the new biotechnology firms have called on a wide array of funding mechanisms. Among the most important of these have been investments from venture capital firms (institutional venture capital) and from established companies interested in biotechnology (corporate venture capital). The investments from the latter have generally taken two forms: equity investments and joint ventures. Equity investments, in which established companies buy portions of new biotechnology firms, have enabled the former to keep abreast of developments in the field, perhaps to gauge the best time to enter the field themselves. Joint ventures, in contrast, usually involve a more active combination of R&D contracts and product licensing agreements. Under the terms of these agreements, an established firm often handles the regulatory approval, manufacturing, and marketing of a product after the small firm has done the initial development. The small firm receives royalties from the sale of the product and usually retains the patent on the products (Schoemaker, 1986).

INITIAL PUBLIC OFFERINGS

Another major source of funding for the new biotechnology firms has been the stock market. In the early 1980s, several start-up biotechnology firms set Wall Street records when they first went public. In 1980, Genentech's stock underwent the most rapid price increase in the market's history, climbing from $35 to $89 per share in its first 20 minutes of trading. A few months later, Cetus raised the then largest amount of money with an initial public offering—$110 million (Schoemaker, 1986). Disillusion then set in as results did not match optimistic predictions. Recently, sentiment has improved dramatically. Since October 1985, the biotechnology share index, made up of 60 publicly-quoted companies, has risen 75 percent. This has more than made up ground lost during the years of disenchantment (*Economist*, 1986).

Two events helped prompt the revival of investor enthusiasm. First, takeover bidders were sniffing around biotechnology companies. Second, in October 1985, Genentech, the world's biggest biotechnology company, won regulatory approval in the United States for human growth hormone, which combats dwarfism. It is the first product Genentech plans to sell itself, rather than licensing another company to market it. This is a landmark in the history of commercial biotechnology (*Economist*, 1986).

A breakdown of venture capital-backed Initial Public Offerings (IPOs) by industry is shown in Table 1. In 1985, medical and health related companies accounted for 18 percent of the total number of venture capital-backed companies completing an IPO. Genetic engineering companies did not participate in IPO activity in 1985.

The shift in industry mix over time is dramatic. For example, genetic engineering, which accounted for 11 percent of venture-backed IPOs in 1982, has accounted for increasingly smaller percentages in each of the following years.

Table 1. Venture-backed IPOs by industry, 1981-1985. (Percentage of Companies)

	1981	1982	1983	1984	1985
Commercial Communications	1%	-%	-%	-%	2%
Telephone & Data Communications	6	7	6	13	9
Computer Hardware & Systems	22	26	27	19	18
Other Electronics Related	19	11	11	13	2
Genetic Engineering	6	11	7	2	-
Medical/Health Related	7	23	17	11	18
Energy Related	10	-	2	-	-
Industrial Automation	-	-	2	2	2
Industrial Machinery & Equipment	5	4	1	4	15
Consumer Related	10	4	1	4	24
Other	9	7	9	4	6
Total	100%	100%	100%	100%	100%

Source: *Venture Capital Journal*, February, 1986, p. 13.

WHY GO PUBLIC?

There are several reasons why management takes a company public. The primary reason is to generate new cash for the company and provide liquidity for investors, including insiders. Going public at the right time also provides maximum cash for minimum dilution of equity ownership by the existing investors.

Timing of an issue is very important. An illustration of timing is presented in a study done by Howard (1973). During a 22-month period, (through October 1972), 452 venture companies filed initial registration statements. Of the 234 ventures who filed registrations in 1971, 97 had gone public by October 1972, a period in which the public markets were good and the Dow Jones Industrial index exceeded 900.

The successes of 1972 led more companies to retain underwriters for offerings of securities in 1973. Through October 1973, 90 ventures filed registrations. Only six were successful. Another six were effective, but the best-efforts underwriting was not completed. Of the remaining firms under study, 16 withdrew their registration and 62 remained in registration, hoping for a new issue market rally.

RESEARCH AND DEVELOPMENT LIMITED PARTNERSHIPS (RDLPs)

A source of financing that has rivaled the stock market in size is a type of investment known as an R&D limited partnership. This allows individuals or organizations to invest in a company's research and development and to write off that money as expenses. The investors become limited partners and are entitled to receive royalty payments from future sales of products. Part or all of these royalties are in turn taxed as capital gains, offering an added attraction to this kind of investment.

Since their inception, more RDLPs have been devoted to biotechnology than any other technology segment. According to a draft of a new study that the New York University (NYU) Center for Science and Technology Policy carried out for the National Science Foundation, through the end of last year RDLPs in biotechnology raised $663 million—fully 27 percent of the $2.5 billion total raised through RDLPs (Klausner, 1986). As a comparison, the study, "RDLPs and Their Significance for Innovation," notes that biotech has attracted only about 3 percent of all venture capital support. The report estimates that 85 percent of all funds raised by RDLPs actually goes to research, and that this financing vehicle has sponsored about $0.5 billion in R&D annually since 1982 (Klausner, 1986).

The bottom line is that RDLPs are here to stay for biotechnology. The NYU Center draft concludes, "A company with a solid reputation and impressive track record still has little trouble in attracting funds dedicated to R&D programs" (Klausner, 1986).

A contrarian view of RDLPs was raised by Klausner (1985), when he stated that the market for RDLPs had all but dried up. Moreover, the 1986 tax reform legislation calls for limitations in tax shelters as offsets to tax liability.

OTHER SOURCES OF FINANCING

The new biotechnology firms also have a number of other sources of capital, including interest from funds previously raised, short-term loans, industrial revenue bonds, and equipment leasing. Through these and other funding mechanisms, the new biotechnology firms have generally been able to bring in enough revenue to remain viable, even though many of them have not yet generated actual products for the marketplace (Schoeman, 1986).

Some biotechnology firms have tried to lessen their reliance on R&D contracts and licensing agreements with large U.S. firms to retain more control over the uses and profits of their ventures with foreign companies. In these cases, the start-up firms often retain the rights to sell their products within the United States while selling the overseas sales rights to their foreign partners. In turn, the start-up firms supply either the products or the technology to make the products to the foreign companies. Many observers have questioned the wisdom of this transfer of technology, claiming that in the long run the spread of know-how generated in the United States to other countries will enhance the competitiveness of foreign firms. Most of the new biotechnology companies have deemed the short-term benefits of such an arrangement to be more important than the long-term disadvantages (Schoemaker, 1986).

UNIVERSITY—INDUSTRY RELATIONSHIPS

Another creative financing technique employed by biotechnology companies is the utilization of university talent and research grants to incubate products for potential commercial applications. Recognizing that many states have different programs and approaches to this concept, the author has chosen to examine the state of Florida's efforts in this regard. The policy statement of the Florida High Technology and Industry Council entitled "High Technology and Economic Development—A Technology

Innovation and Commercialization Policy Statement" (1985) begins by describing a long-range economic development program for Florida that targets science and high technology industries as key markets, and stresses the importance of working relationships among education, government, business and industry to maximize economic development opportunities. The paper describes two additional development strategies being proposed: spin-off industries and innovation industries. Spin-off industries are the result of expansion and/or diversification by existing industries. Innovation industries are the result of new industry development based on scientific discovery and technological innovation. One of the key components in this concept is Florida's educational system, which must provide both the idea generators (via the university system), and the idea implementors (via the vocational technical schools). As the generator of high technology innovation, the university system is shown to be the cornerstone of the economic development process. This document also stresses the point of view that "business, particularly small entrepreneurs, needs a steady, reliable stream of risk capital. Encouraging support of new business by the investing public is identified as a key factor of development success. Recent state initiatives to fund technology start-ups provide a partial solution."

The key issue for the state is how to expand the working relationship between industry and the universities with special attention directed towards (Florida, 1985-a):

1. Removing the barriers to collaboration between university faculty and industry.
2. Improving university research programs, particularly in areas with potential economic return.
3. Providing incentive mechanisms to promote a broad range of cooperation and collaboration between universities and industry with specific emphasis on support for research efforts with potential economic benefits.
4. Improving the availability of risk capital. A particular problem related to university research is the availability of capital for the initial, high-risk components of the technology transfer process.

Over the past few years, Florida has made significant strides in addressing each of these issues.

Biotechnology is extremely dependent on the universities, perhaps more so than other high tech industries. The plan for developing a biotechnology industry for Florida has three major components (Florida, 1985-b):

1. The enhancement of those emerging biotechnology strengths at the

four major research institutions that coincide with the needs of Florida for which this technology holds promise.
2. The development of a competitive grants program that promotes innovative approaches in research and training that particularly address Florida's needs and which is accessible to all Florida scientists regardless of affiliation.
3. The establishment of a technology transfer center whose mission is to serve industry, the universities, and the state in a dynamic integrated fashion that very effectively pools Florida's resources and, through its training programs for technicians and bioprocess technologists, provides an immediate means for Florida to assume a leadership role in biotechnology world-wide.

In August, 1986, the Governor's High Technology and Industry Council launched a contest in the scientific community in the state. The challenge was to submit a research proposal to establish Florida as a leader in the development of one of six technologies, which includes biotechnology. Winners would share the proceeds of a fund the Legislature would replenish annually for five years. Governor Bob Graham's goal is to nurture and speed up the new technology from the laboratory to the marketplace where it can be converted into new Florida companies and more Florida jobs (Selz, 1986).

Robert F. Johnston's (1986) paper, "The New Connection," describes a new relationship between the university, the small entrepreneurial company and the large corporation. "The university is a source of basic research technology; the entrepreneurial company sees the need in the marketplace for specific products that can be made by applying some of the basic technology from the university; and the small company enters into an arrangement with the larger corporation for marketing, manufacturing and, possibly, financing." Johnston continues by stating that "in areas such as biotechnology, where there is a high rate of technological change, the competition is keen, there is great pressure for the small company to identify niches in the marketplace and to move very rapidly to capitalize on these. The entrepreneurial companies have the resources and ability to do this. In my experience, they do, in fact, recognize the opportunities, structure contracts and research relationships with outstanding members of the academic community able to support them on the project, and then select a larger corporation with the marketing skills to distribute the product." Johnston feels that his plan contains benefits for all. "The university retains its talent pool, the small company cultivates the resources of the university and the larger corporation maximizes on its established position in the marketplace to manufacture and obtain regulatory approval of the product...To summarize, all three parties should benefit from this collaboration. The university receives additional research funding, the small company speeds the commercialization of the new technology and the large company has a new product to manufacture and market." He sees this relationship continuing very effectively in today's environment.

In "High Technology Industry in Florida: Approaching Critical Mass" (Gonzalez *et al.*, 1985), the authors cite a study conducted by SRI International entitled "U.S. Government Programs and Their Influence on Silicon Valley." The latter report states it was determined that factors for success other than the role of government included "the existence of a unique industry-university relationship...It is clear that such an important relationship exists in both California and Massachusetts, where the success of Silicon Valley and the Route 128 area is closely tied to the interaction between industry and strong universities. In an attempt to stimulate this relationship within Florida, the state is encouraging the development of science and high-technology research parks. Presently, four university-related research parks have been established in Florida. The state also has established over 200 special research centers at various universities."

The authors conclude by outlining a number of important issues that will become critical determinants of the future success of Florida as a high-tech state. First, Florida must continue to take a unified approach in encouraging and the locating of major high-tech companies within the state.

Florida's university system has not yet established the type of relationship with industry necessary to foster the kind of explosive growth that occurred in California and Massachusetts. This type of growth is likely to be driven by the industry innovators, and not by the university system as many believe. Universities respond to, but generally do not create, industrial revolutions (as contrasted to scientific revolutions).

Finally, the ability of the state to attract and support the entrepreneurially minded also has tremendous potential impact. Creating the proper atmosphere for entrepreneurial risk taking cannot be over emphasized.

Ultimately, the fate of high tech within the state will not rest merely on a pleasant lifestyle, a good university system or a favorable tax structure but rather on those leaders with a vision of Florida's high-tech future who will take the extra risks and make the extra efforts to bring Florida into the forefront of the high-technology industry, and into the 21st century (Gonzalez *et al.*, 1985).

While the foregoing citations on industry-university relationships are very positive, there are those who state that the benefits are not as positive.

Miller and Côte (1985) state that despite these legitimate roles for government and universities, most government laboratories and universities are poor incubators of entrepreneurs and high-tech products. Usually, neither their researchers nor their laboratories have any significant contact with the marketplace. A potential entrepreneur working in such an institution is not exposed to the market, its needs, its organization or its people. Product ideas generated in government laboratories and universities seldom meet marketplace standards either technologically or in terms of cost. They go on to state that the ideal research environment to support a high-tech cluster has the following characteristics:

- The region has several research institutions, like research oriented universities and laboratories, which are recognized as leaders in their fields and boast a significant reputation to attract "the best and the brightest."
- A tradition of contract research exists in these institutions.
- A few large corporations have set up advanced laboratories in the regions where they conduct their basic and generic research.
- A tradition of close relationships between those research institutions and local high-tech companies has taken root through consulting contracts, hiring of graduate students, and occasional joint venturing.

In summary, the pursuit of government grants, university research contracts, and other forms of similar creative financing have allowed biotechnology companies to pursue their development plans. Much of this financing is necessary when more traditional avenues of funding are unavailable to new firms.

TRADITIONAL REQUIREMENTS FOR A SUCCESSFUL INVESTMENT

While the financing vehicles may be changing, the reasons behind the decision to invest or not remain largely the same as they have for many years. These reasons are categorized in three general types: external environment—"The Market," company specific issues—"The Opportunity," and management—"The key" (financing, previously discussed, being the fourth).

THE MARKET

According to the *Economist Newsweekly Industry Brief* (1986), "by 1984, initial disappointment had prompted about four-fifths of America's 200 biotechnology firms to change course. They went from research/genetic engineering of plants and the production of bio-food to sell diagnostics and pharmaceuticals to the fast-growing health-care industry. They reckoned they had an edge in this market sector as their products were based on technologies which were then inaccessible to conventional drug houses."

An example of an external factor follows. "Monoclonal antibodies provided the first products of the biotechnology industry. This market has become intensively competitive. The technology to make the antibodies has also become so cheap and easy that latecomers, including many long-established drug companies, are now developing rival products. Forecasts made by industry analysts a few years ago put sales of diagnostics at around $500 million worldwide by 1985, rising to $2 billion by 1990. In fact, in 1985, biotechnology firms sold only about $150 million worth, partly because they could not take account of cost-cutting measures imposed on American hospitals under the Reagan administration, but mainly because they were too optimistic. Nor do prospects for diagnostics look bright. Regulatory requirements could become stricter. The American Food and Drug Administration wants to introduce tests of efficacy for cancer diagnostics. If it gets its way, it might take ten years to show whether a diagnostic can successfully prevent a full-blown cancer."

Patents present another hurdle for biotechnology companies. Only a few patents have been awarded on drugs made by genetically engineered bacteria. Even if patents are made to stick, other barriers block the maximization of profits. American export restrictions prohibit a manufacturer from exporting drugs which have not received FDA approval (*Economist*, 1986).

Yet another external factor looms large in the biotechnology industry, Mr. Jeremy Rifkin. Mr. Rifkin, a professional environmentalist, who has fought almost single-handedly to hold back the biotechnology revolution, is attacked by the industry and some scientists as hysterical, shallow, and anti-intellectual (*Wall Street Journal*, 1986). He pays an almost loving attention to the details of bureaucratic procedures in an industry where such procedures have historically been fuzzy, over-lapping or downright confused. To gain this national soapbox, Mr. Rifkin challenges biotech experiments on procedural grounds. There is tremendous negative feeling on his actions within the scientific community; however, financiers must take into account long, procedural delays for testing products because of his actions. Regardless of their merits, those procedural acts place tremendous strain on a company's cash flow and ability to commercially exploit a product.

In summary, the market is a key to success. The company must understand the dynamics of the marketplace in which it competes. Knowledge of the strengths and weaknesses of competitors and their products is necessary. Of vital importance is how its products can replace other products currently being purchased. The market must be of sufficient size and internal growth potential to allow company success. This analysis of competitors must be sufficiently clear so that venture capitalists can be convinced that the new company will be able to function effectively and out-compete existing products.

THE OPPORTUNITY

Although products are not yet flooding the market, and juicy dividend checks are not being sent out to investors, the scientific-business community increasingly is talking about biotech's "maturation." Tell-tale words like "manufacturing," "marketing," and even "consolidation" have started to join "promising" and "research" in press releases from biotech companies. Even the buzzword "vertical integration" is beginning to have some meaning for biotech. Manufacture, on one end of the production chain, and distribution on the other are becoming real concerns as firms prepare to bring products to market or at least supply their wares for clinical trials.

Distinct corporate strategies are emerging, a development that renders the start-ups easier to distinguish—much to Wall Street's relief. Companies are saying to themselves, "What do I want to be when I grow up?", points out Patricia D. Berkley of Arthur D. Little's applied bioscience unit. Genentech, for one, has been claiming that it is or will soon be a pharmaceutical company (Klausner, 1985).

MANAGEMENT

A critical problem facing new biotechnology companies is managing rapid growth and change. Not only are biotech companies highly volatile in their early years, but they operate in a marketplace that is complicating things by changing rapidly. "Management at biotech companies must understand this growth process, because crisis is endemic to it. The company that can anticipate and manage crisis—rather than simply react to it—is in the best position to turn it into a growth opportunity" (Burill, 1985).

Steven Brandt of Stanford University and Larry Greiner also have looked hard at what's happening to growth organizations. A discussion of their ideas and my firm's experiences in serving the biotech world can demonstrate how companies that understand challenges and crises can maximize profitability and grow immeasurable.

The keys to success lie within a company itself. External events—such as weak markets—matter, but how a company responds to these events is what determines whether the company moves forward.

A.D. LITTLE'S SIX PHASES OF CORPORATE DEVELOPMENT, AS INTERPRETED BY BURILL

1. The start-up

"One important window for biotech companies in this start-up growth stage is the opportunity to develop credible business plans. Many companies regard this plan merely as a document. It is not. A business plan is a process. It is guide to corporate growth."

"During the start-up, a company's unique corporate culture is defined. The founders send clear signals to the employees and through these signals they set a management style and philosophy. Because building the team is so critical in this early stage, it is important to let others know that retaining talented people and teamwork are priorities in the culture. Many biotechnology firms spring from the inspirations and outstanding research of first-rate scientists. But as companies grow, they must understand that a well-balanced team of professionals is far more valuable than an assortment of individual stars."

"Moreover, good scientists do not always make good business people, and companies formed to profit from basic scientific research must acknowledge the need to eventually hire professional management staff. Without balance and teamwork among scientific, finance, production, and marketing people, the chances of eventual profitability are greatly diminished."

2. Nurturing creativity

"One of the biggest problems in going forward is focusing on an area that offers the best chance of success: identifying the market that is largest and can be penetrated most quickly and setting up appropriate milestones to effectively manage through that market entry."

"This second phase can be particularly stressful. We've seen that finance strategy usually will dictate business strategies—not vice versa, as many think. It also requires focusing on short-term investor expectations rather than long-term business needs."

3. Management and control

"These milestones can cause leadership crisis. The company's founders and senior managers may now be burdened with unwanted or unfamiliar management responsibilities, yet they may be reluctant to step aside. To resolve this crisis, the company frequently needs to bring in capable business operational managers familiar with these new demands."

"Rapid growth requires the rapid development of a corporate structure. But as this structure takes shape, communication between management and staff must be maintained."

4. The move toward delegation

"As companies become larger they often encounter an autonomy-related crisis, launching them into their fourth growth phase."

"Obtaining the right people to lead the company becomes a balancing act. The company wants to attract talented entrepreneurial people who would not work for conventional corporations (where incentives are too indirect), but matching large organizational skills with incentive-driven people is difficult."

5. Growth through coordination

"Inherent at this time is a crisis of control...A management tier strong in personal leadership skills—the ability to get things done through other people—is of greater importance than executive management experience."

6. The collaborative approach

"Biotech companies at this point can become buried in red tape...This can be overcome by an approach based on collaboration: mature companies use team approaches to problem solving."

THE ENTREPRENEUR

We have examined the phases of a biotech company's development in the preceding section. Are biotech firms different from other venture-capital backed firms? They are not. Biotech firms need the vision of an entrepreneur to get things rolling.

The entrepreneur plays a significant role in the progress and development of a technology business. Venture capitalists usually look for a person with management experience and profit and loss responsibility who is striving to advance himself in any honest way. Frederick Adler (1973), a noted venture capitalist, gives his assessment of the ideal entrepreneur as follows:

"He must have the intellectual integrity to admit his mistakes and to recognize and reward other people's talents. This man must be technically qualified to do the job but must not be so immersed in the technology that he loses sight of the need to build a profitable business rather than a bunch of fancy products. He must be a man of ego. If he does not have a very large ego, he is not going to view the obstacles with sufficient confidence. Yet, if he is too egocentric and one-sided, he will make some serious, dumb decisions because he refuses to take input from others."

"The man must be tough enough to make very hard decisions if the venture is to survive, and that means firing his best friend if necessary. Yet, he must be smart and mature enough so that his toughness is tempered so that the people he needs around him will not leave because his attitudes irritate them."

Harry Schrage (1965) undertook a research project at Massachusetts Institute of Technology to determine what makes a successful R&D entrepreneur. He interviewed 22 R&D president-founders regarding their attitudes, decision-making methods, and management problems. Schrage found that the ability of the entrepreneur to perceive the reactions of the market to his firm's product or service, and to be acutely aware of his employees' morale, directly related to high profitability. He suggests that the following questions ought to be asked by business associates, suppliers, bankers, and others in dealing with the R&D entrepreneur:

(1) "Does he appear to really know the nature of his customers? Is he aware of why they buy his products or services? Does he know what they like and dislike about his company? Above all, does he appear to have made an active effort to find out?"

(2) "Does he really know his employees? Can he distinguish between the more-and-less-effective ones? Or does he stereotype all his employees with one brief description? Can he distinguish several factors that make for good morale and others that tend to destroy it? And again, has he taken the trouble to find out for himself, or does he just appear to be theorizing?"

(3) "How curious is he about people? Indications of curiosity are good signs of awareness."

(4) "Does he tend to look up and compare his company to highly successful ones, or does he look down to failures for comparison? An as yet unreported characteristic of the highly successful entrepreneur is the tendency to be critical of his own performance."

Schrage summarized the successful entrepreneur as being "high in achievement motivation, low in power motivation and high in awareness of self, the market and his employees."

Another view of the importance of management in technology-based businesses is expressed by Rind (1973). An entrepreneur has to be aware of the fact that the management team really is the business, and that the quality of management is all important. The success of a venture investment in a technology situation seems to center on a team led by a manager.

A fitting summary to this chapter is expressed by Koch *et al.* (1983). "High-technology companies bring to a region not only new jobs, but also new ways of managing people to produce a product in a highly competitive environment. Our research brought to light a quiet revolution going on in American industry. Those companies on the cutting edge of behavioral change. Successful high-technology companies are led by enlightened managements. They are disillusioned with traditional corporate structures. They believe there must be a better way to operate a business. The model they provide is one of integrating people with technology to get results. The ingredients necessary for successful operation of a technology firm are increasingly becoming the same ingredients needed to operate successfully in traditional industries. High technology is a product of the information age. A management style conducive to this age naturally has developed."

CONCLUSIONS

The biotechnology industry and the companies participating in it have experienced a roller-coaster ride. Their great expectations of the late 1970s reached a crescendo of optimism with the first public offerings of many of the firms. "Going public" perhaps raised false expectations of quick profits to the investing public. The resultant disappointment as product development and commercialization dragged on was to be expected. As products are finally certified for use, the investor enthusiasm is quickly re-kindled.

This is not an unusual environment for a venture capitalist. The biotechnology issue that has concerned financiers is the very lengthy, and entirely uncertain paybacks from the projects. External factors, such as legislation, regulation, and glutted markets exacerbate this problem.

As traditional avenues of venture capital have been shut off to developing firms, creativity played an important role in raising financing. Research and development partnership and grant financings, licensing agreements and technology transfer arrangements arose to supply much needed funding to continue development efforts.

The allure of biotechnology to investors and corporate partners is not universal. Finding compatible partners for business arrangements is critical to commercialization of biotechnology.

The investment in a biotechnology company is not considered to be different from other technology ventures. Furthermore, the tried and true axioms of financing apply fully to any business. Entrepreneurs with a commercial vision, a realistic business plan, a business opportunity, and a robust market for the product are keys to success. Expert management then continues the maturation of the idea into corporate reality.

REFERENCES

Adler, F. (1973). The art of venturing. In *Guide to Venture Capital Sources* (S. Rubel, ed.), Third edition, p. 71. Capital Publishing Corporation, Chicago.

Bailey, S. (1978), "Financing the Independent Computer Peripheral Manufacturers: A Case for Industry Specialization," unpublished thesis submitted in partial fulfillment of the requirements for graduation from the Stonier Graduate School of Banking, conducted by The American Bankers Association at Rutgers University, New Brunswick, New Jersey.

Bailey, S. (1979), Lending to the manufacturer of computer peripheral equipment. *The Journal of Commercial Bank Lending*, vol. 61, no. 10, June, 1979, pp. 18-31.

Burill, G., (1985). Patterns of growth. *Bio/technology*, v. 3, No. 10, October, 1985, pp. 875-879.

Economist (1986), Biotechnology's hype and hubris. v. 299, no. 7442, April 19, 1986, pp. 96-97.

Fitzsimons, J. (1984), Do's and don'ts in attracting venture capital. In *Raising Capital or Emerging Companies*, (Deloitte Haskins & Sells—seminar participants' notebook), Tampa, Fl.

Florida High Technology and Industry Council (1985). *High Technology and Economic Development: A Technology Innovation and Commercialization Policy Statement*. pp. i-iii, 9-10, 13, 29. Executive Office of the Governor, Tallahassee, FL.

Florida High Technology and Industry Council (1985), *Report to Science Panel on Biotechnology*, February, 1985, pp. 4-6, 14.

Gonzalez, J.; Maidique, M.; and Susan, L. (1985). *High Technology Industry in Florida: Approaching Critical Mass*, pp. 17-18, 45-47. Innovation and Entrepreneurship Institute, School of Business Administration, University of Miami, Coral Gables, FL.

Howard, G. (1973). Going public when it makes sense. In *Guide to Venture Capital Sources* (S. Rubel, ed.) Third edition pp. 74-75. Capital Publishing Corporation, Chicago, IL.

Johnston, R. (1986). The New Connection, an unpublished paper, pp. 1-4.

Klausner, A. (1985). Biotech's first steps into the business world. *Bio/Technology*, v. 3, no. 10, October, 1985, pp. 869-872.

Klausner, A. (1986). R&D Limited Partnerships Start to Pay Off. *Bio/ Technology*, v. 4, no. 4, April, 1986, pp. 268-271.

Koch, D.; Cox, W.; Steinhauser, D., and Whigham, P. (1983). High technology: the southeast reaches out for growth industry. *Economic Review*, September 1983, The Federal Reserve Bank of Atlanta, Atlanta, GA.

Miller, R. and Côte, M. (1985). Growing the next Silicon Valley. *Harvard Business Review*, July-August 1985, p. 116.

Murray, J. (1986). The first $4 billion is the hardest. *Bio/Technology*, v. 4, no. 4, April 1986, p. 293.

Rind, K. (1973). The potential and problems in financing technological companies. In *Guide to Venture Capital Sources* (S. Rubel, ed.), Third edition, p. 34. Capital Publishing Corporation, Chicago, IL.

Rubel, S. (1973). Raising capital for small companies: an overview. In *Guide to Venture Capital Sources* (S. Rubel, ed.) Third edition, p. 10, 26, 28. Capital Publishing Corporation, Chicago. See also, *Pratt's Guide to Venture Capital Sources* (S. Pratt, ed.) Published by the Successor to Capital Publishing Corp. Venture Econmomics, Inc. Wellesley Hills, Mass., 10th ed.

Schoemaker, H. (1986). The new biotechnology firms. In *Biotechnology: An Industry Comes of Age* (S. Olson, ed., for the Academy Industry Program of the National Academy of Sciences), pp. 84-87. National Academy Press, Washington, D.C.

Selz, M. (1986). A blue ribbon research team takes to the marketplace. *Florida Trend*, v. 29, no. 1, May, 1986, pp. 69-73.

U.S. Congress, Office of Technology Assessment. (1984). *Commercial Biotechnology: An International Analysis*. Superintendent of Documents, Washington, D.C.

The Wall Street Journal, 1986. Jeremy Rifkin usually infuriates—and often beats—biotech industry. May 2, 1986, p. 23.

Commercializing New Products from Plant Biotechnology: Problems and Perspectives

Cyrus McKell
NPI, Salt Lake City, Utah

INTRODUCTION

Possibly the most important phase of biotechnology research and development is product commercialization. Obviously, all management functions must be successfully carried out to encourage and support product development. Teamwork is essential at the corporate level, as well as in all aspects of management, to ensure effective and timely use of resources and provide the proper environment for the elaboration of products that have been promised. This is not to say that the dearth of products up to the present time indicates that the industry lacks good management. Indeed, the challenges assumed and the obstacles facing product development are formidable, and only the companies that have good management and research will be able to succeed.

Plant biotechnology companies are generally newer than their counterparts in the fields of human health, veterinary medicine, and pharmaceuticals. Thus, the appearance of plant products has lagged. However, plant biotechnology has the advantage of utilizing previous research that provides the basis for biotechnology in general, but can be applied specifically to solve problems in plant agriculture. Although we have seen precious few results from plant biotechnology during the past two or three years, we now can expect to see new products and revolutionary processes tumble out of this scientific cornucopia that will bring benefits to an agricultural industry faced with the necessity to increase its efficiency and be more innovative.

Tissue culture and various related procedures were the first modern methods of plant biotechnology to be applied to create new products. Cloning of cymbidium orchid shoot apices in 1960, by George Morel in France, allowed the mass production of selected high value plants of identical genotype. Subsequently many orchid growers around the world adopted tissue culture techniques and by 1977 (Arditti, 1977) 35 genera and intergeneric hybrids had been successfully cloned for commercial production. Unique trees and shrubs (Bonga and Durzan, 1982) can be multiplied to yield many more copies than could be produced by conventional techniques of grafting or rooting of cuttings. At NPI we have supplied a number of trees and shrubs to our nurseries from tissue culture. We are also using *in vitro* culture techniques to produce precocious flowering roses marketed under the name "Forever Roses" that are intended as house plants in the winter but which can be planted outside in the spring (Bylinsky, 1985). Aseptic techniques for tissue culture production provide a means for the multiplication of disease-free potato plantlets to produce the "nuclear" generation which can be planted for subsequent field multiplication of seed potatoes (Upham, 1982). By selecting disease-resistant sugarcane plants and then cloning them via tissue culture embryogenesis, Crop Genetics International has been able to provide plants that afford growers a yield increase of 20 percent (Bronson, 1985). In addition to the production of plants, numerous strategies for improving plant agriculture are approaching the patent stage such as biological pest control, herbicide, and stress-resistant plants, and development of the male sterility factor for use in breeding plant hybrids (Lawton, 1986).

The purpose of this paper is to discuss some of the problems and challenges facing the plant biotechnology industry as it gains experience in commercializing new products. Techniques of management will be discussed in relation to creating an innovative environment for product-oriented research. An example of a new product that is in the early stages of market testing at NPI will be used to illustrate some of the points of the discussion.

TARGETING RESEARCH TOWARD NEW PRODUCTS

Whereas university research seeks to publish results from investigation of basic biological principles, as well as ways to apply principles to solve problems, commercial biotechnology research has the goal of developing products or patentable processes based on research from all available sources. Another distinction is that results of research sponsored by the State Agricultural Experiment Stations generally find their way to the user group through Agricultural Extension. Contrasting this, results of industrial plant biotechnology research must be translated into products that can be marketed at an affordable price—often to farmers but also to agricultural industries, and even home gardeners. Clearly, products are the vehicle of communication from the plant biotechnology industry to society rather than information. Thus, biotechnology further shortens the linkage between research and its users.

APPLICATION OF BIOLOGICAL PRINCIPLES

Making a clear distinction between basic and applied research in biotechnology is somewhat meaningless because the goal must be to apply whatever knowledge is available to product development. Often it is necessary to conduct fundamental research that continues into applied studies in order to sufficiently manipulate biological systems. Ideally, fundamental knowledge necessary to understand such systems already exists in the scientific literature and all that is necessary is to adapt the principles to solve the problem at hand. The old saying that "knowledge is power" is particularly apt in finding opportunities to apply biological principles. Scientists in the biotechnological industry have before them a wide array of opportunities to manipulate biological systems and genetic materials, but the target must always be to work toward solving problems that will culminate in marketable products. Our method at NPI for targeting research is to elaborate our research ideas in a proposal focusing on the end product.

SOLVING AGRICULTURAL PROBLEMS

Clear identification of the agricultural problems that can be solved by biotechnology is one of the major challenges to the scientist and research manager. We have three basic options in solving agricultural problems: increase yield, improve quality, or reduce production costs. In view of the large surpluses in some crop commodities, the best options would appear to be the latter two. Biotechnology can now solve problems that defied conventional agricultural research in the past. For example, by using tissue culture techniques, it is now possible to select among millions of cells for tolerance to salinity, drought, disease, pesticides, and other stresses. When a more specific approach is needed, we may consider transferring the gene(s) that control the genetic mechanism conferring stress tolerance.

Given the capability to solve an infinitely large number of problems, plant biotechnologists must target the type of research to undertake as well as the crop species to be improved. Several criteria may be used in making the selection of the technology, such as the nature of the crop, the market value of the crop, the benefit to society, or the degree to which the problem would be solved, i.e., producing disease-free plants or the more complicated and long-term objective of creating disease-resistant plants.

MARKET JUSTIFICATION

The cost of the research in relation to the value of the expected product is undoubtedly the most important consideration in selecting the research target. Heretofore, the issue of cost has not been as important in the funding of research through the Agricultural Experiment Stations as it now is in the plant biotechnology industry. Projects were funded on the basis of the benefit to society regardless of the need to capture income from the research results. As a commitment to investors who have placed over $4 billion in the biotechnology industry (Murray, 1986), research must provide a return on the investments. For example, some may ask whether or not we should improve crops such as cassava that serves as a food staple for millions of people in many tropical countries. This crop is propagated vegetatively as a non-market food source and is not distributed extensively in the market. Thus, even though there may be a compelling need to improve cassava productivity and quality, there is little financial incentive to do so and little opportunity to distribute improved genetic material to the users. At present, the major commodity crops such as corn, tomatoes, rice, potatoes, sugar cane, ornamentals, and vegetables that are traded as proprietary materials with a large market value are the most attractive targets of plant biotechnology research. After methodologies have been worked out to improve the major crops and increase the effectiveness of their culture practices, we may be able to apply biotechnology at a lower cost than initially to lower value crops or to those where less proprietary control exists.

INNOVATIVE RESEARCH: CREATING THE ENVIRONMENT

The biotechnology industry must foster a creative research environment. Scientific staff like to feel they have opportunities to grow professionally, that they are not in scientific isolation because of being in industry, and that their ideas are important and worthy of testing. Because of the high regard research staff hold for the academic community through role models relating back to their university training, management should try to create an environment that simulates and builds upon the academic experience of the scientific staff.

UNIVERSITY COOPERATION

Recognizing the essential linkage between basic research and its application to products, biotechnology has forged important cooperative relationships with universities. The relationships take on many forms such as cooperative programs funded by major grants, joint research projects for university-industry studies awarded by the National Science Foundation, and private consultantships with senior university scientists. Industry often has an opportunity to participate in university activity through adjunct professorships and presentation of seminars. Exchange of ideas at professional meetings among university and industry scientists is vital to innovative research. In such interchanges, university scientists receive feedback from industry as to problems for solution and industry scientists learn about new hypotheses being formulated which may have long-term implications for product application.

In the early development of research at NPI, a Technical Advisory Board was organized to review the research program and suggest opportunities for products. However, as the research effort grew in diversity and strength, the need for general advisory board changed. We presently appoint technical review consultants who review programs and meet with research teams to stimulate innovative ideas. Most of these consultants are from leading universities and thus provide an additional linkage between NPI and the scientific community.

MULTIDISCIPLINARY TEAMWORK

Another stimulus for innovation comes from the interaction of scientists from diverse disciplines working together as team members. Lacking the departmental or college boundaries that sometimes deter close teamwork in a university setting, the biotechnology industry is able to create an innovative environment for interdisciplinary teamwork. An example of this is the development of an insect-resistant tomato (Nienhuis *et al.*, 1985), in which a plant geneticist, a molecular biologist, and a plant products chemist organized a team to determine the inheritance pattern of the biologically active chemical, 2-tridecanone, to use restriction fragment length polymorphisms (RFLPs) as molecular markers in crosses between *Lycopersicon esculentum* and a wild species, *L. hirsutum f. glabrum*, and to develop a chemical test that could serve as a quick index to the quantity of the chemical present in plants. Similarly, most research and development in biotechnology is multifaceted and requires the input of several scientific specialties.

SUPPORT SYSTEMS

Adequate equipment, laboratory facilities, and research greenhouses are essential for research success. Management must ensure that the workplace and its equipment will foster an effective pursuit of research objectives in safety. Obsolete and time-wasting equipment should continually be replaced. Scientists need to feel that management is concerned about effective use of both time and resources.

Staff management should be positive and sufficiently open to encourage interaction and innovation, striving at all times to avoid developing "territorial boundaries." At NPI, research divisions are under the leadership of a coordinator whose task it is to facilitate day-to-day activities, coordinate between the director of research and the staff, and promote effective laboratory operations. In contrast with this open management, multidisciplinary projects are under the control of a project manager whose task is to see that project goals are met within the allotted time while making use of all resources available to the project.

STAFF DEVELOPMENT

The most valuable asset of a biotechnology company, in its pursuit of products, is its staff. Continued development is essential to creativity. Any worthy activity that promotes scientific growth should be supported. Some of the most productive activities appear to be attendance at scientific meetings and workshops, presentation of seminars, and publication of scientific papers. In-house training activities include work-in-progress reports and staff seminars. Corporate support for participation in scientific meetings and university coursework taken by junior scientific staff can add new technical "tools" and sharpen existing staff capabilities, but it also sends a message to the staff that continued scientific growth is a high priority.

NURTURING INNOVATIVE IDEAS

A unique characteristic of research is that alternative answers and solutions to problems can be applied to new product areas through innovative thinking. Many of the successful corporations described by Peters and Waterman (1984) in their book, *In Search of Excellence,* encouraged some form of "bootleg" research during current project development to create and test new ideas and products on an ad hoc basis. Small research and development teams were given free rein, often informally, to sort out new approaches and encourage innovative thinking. Once an idea is shown to be feasible, it needs to have a "champion" within the corporate structure to keep it alive and growing. Some of the most innovative companies encourage senior staff to act as a sponsor of good ideas.

At NPI, we encourage innovative ideas by providing in-house seed money to the most promising proposals. Creative research, "bootlegging," on ideas tangential to budgeted research is encouraged (without stinting on the commitment to the original goals). In all cases where feasibility for new approaches is shown to be high, ad hoc development teams are formed to prepare proposals for indepth research and development of the idea(s). Funding for these new innovative ideas comes from a variety of sources such as in-house budgets, and Small Business Innovative Research (SBIR) grants from federal agencies. Subsequently, university-industry grants from federal research funding sources, research and development limited partnerships, and corporate joint research ventures provide an array of opportunities for sponsorship of innovative research ideas.

EARLY PRODUCT IDENTIFICATION

Although a product or patentable process is the goal of the biotechnology industry, early identification of the nature of the product is necessary to facilitate corporate planning. Development of Nutri-Link, a VA Mycorrhizal inoculant from NPI for horticultural crops, serves as an example of research management to commercialize a product from biotechnology.

PRODUCT EXAMPLE: VA MYCORRHIZAE

Products from biotechnology must be superior to those already in use or else they must provide new ways of improving agricultural production. Nutri-Link fits this second criterion. Vesicular arbuscular (VA) mycorrhizae is a beneficial root fungus that is associated with nearly all plant roots (Gerdemann, 1975). This root symbiont grows into the cortex of a root where it produces structures termed arbuscules and vesicles within the root cells (Figure 1). External to the root, the hyphae proliferate into the soil in considerable abundance and serve as a conduit for uptake of water and plant nutrients as well as form a protective screen against many plant pathogens in the soil (Menge, 1980). One of the first practical applications suggested for mycorrhizae was to inoculate seedlings to assist in their establishment upon being transplanted into sterile soils following massive disturbance from surface mining and in revegetating spent oil shale disposal sites (Aldon, 1978; Call and McKell, 1985). Concurrently, Professor Menge at the University of California, Riverside, demonstrated the use of inoculants on citrus transplants and other horticultural crops to increase their establishment success.

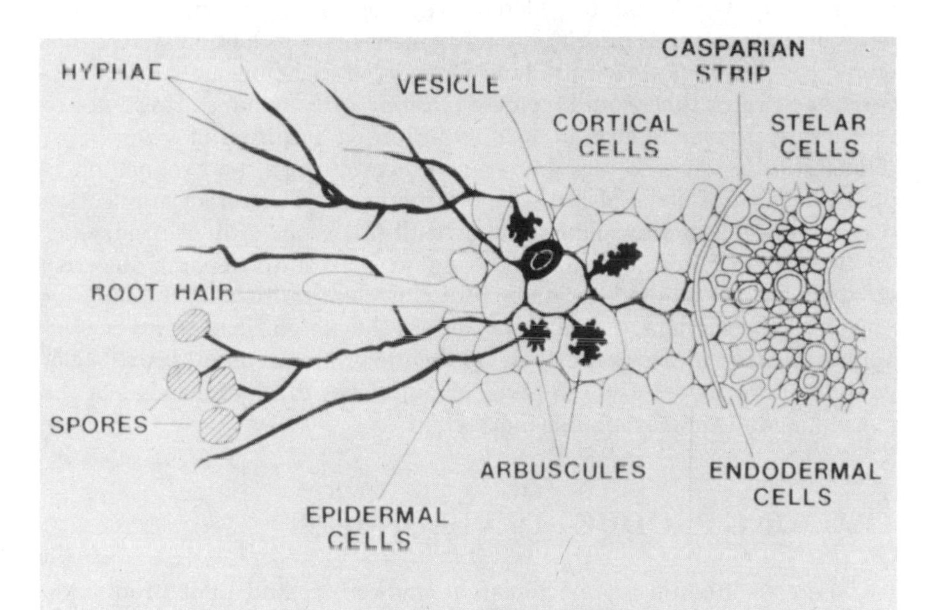

Figure 1. Cross-section of a root showing the relation of mycorrhizal structures in the cortex and external area of the root. The greater extent of hyphae compared with root hairs assist in the uptake of nutrients and water.

The barriers to commercialization of mycorrhizae were the lack of superior strains (even though genetic variation appeared to be sufficient in the populations to warrant a program to make selections), inability to produce the strains in pure culture, and procedures for applying inoculant in sufficient quantity to be effective under conditions of competition with other soil organisms to the degree that the beneficial influence of the VA mycorrhizae would be negated. Obviously, with these obstacles to overcome, research priorities had to be established.

RESEARCH OBJECTIVES

Identification of a product or series of products anticipated from research is essential to formulate research objectives and place them in priority. Very early in the commercialization of VA Mycorrhizae, the project manager, Dr. Tim Wood, considered several options for product areas. He set as his first research priority the refinement of a production system that would provide a consistent product. Concurrently, he began to select mycorrhizae strains that would increase plant growth, enhance resistance to root disease, promote crop uniformity, improve hardiness of transplants, reduce stunting of seedlings grown in fumigated soils, and reduce phosphorus and trace-element fertilizer requirements. The target application was inoculation of transplantable horticultural crops grown in nurseries on fumigated soils or in sterilized potting mixes. A third research objective was to develop a method for mass-producing clean cultures of the superior mycorrhizal selections. Research objectives that might lead to other crop applications, or methods for applying the inoculant on an extensive scale, were temporarily set aside in favor of objectives that would lead to the most immediate and attainable market.

TRACKING PRODUCT DEVELOPMENT

Close coordination among research, marketing, and pilot production was vital to ensure that a sufficient volume of the new product was ready for test-marketing at the critical season. NPI's field tests of VA Mycorrhizae in California provided answers to questions about plant responses and the performance of selected mycorrhizal strains and combinations. As field responses were evaluated, directions were clarified for product formulation, pilot production, and marketing strategies. Patent opportunities now needed to be examined in light of the research results.

PATENT PERSPECTIVES

According to patent attorneys, a patentable invention or plant product must be new, useful, and unobvious (American Bar Association, 1982). In legal terminology, for something to be unobvious, the differences between the subject matter to be patented and the prior item are such that the new item would not be obvious to a person of ordinary skill. The invention must differ from the prior art in a way that is not just an obvious change or addition. The person applying for a patent is the first and original inventor.

The steps outlined to NPI in applying for a patent include: preparing a disclosure of the invention, conducting a patentability search of the invention, preparing and filing a formal patent application in the U.S. Patent and Trademark Office, and prosecuting the application in the patent office.

Preparation of the patent document starts with the written description of invention which includes a review of the technical literature and the concepts underlying the invention. The patentability search consists of a review of all previously filed patent applications (prior art) in the U.S. Patent and Trademark Office in Washington, D.C. If the invention appears to be patentable according to the original criteria, a patent document is prepared which defines the scope of patent protection that is sought. Close coordination between the inventor and the patent lawyer is important. Once the patent application is filed, the invention may be labeled "patent pending." The filing date is important to indicate priority in case a competing application is filed for a similar idea or material. After considerable study, a patent examiner may issue a report called "Examiner's Action." The Examiner's Action may range from a recommendation for issuance of a patent, a request for further information, a request to reduce the scope of the claims submitted, or a finding that the application lacks direction toward a patentable invention. On the average, patent applications currently require more than 18 months.

NPI's patent attorneys advised maintaining consistent and clear documentation of research activities in laboratory notebooks. Dates and details of experiments leading up to the invention or biological creation are vital. Disclosure of the invention and data supporting its validity must be followed by a patent application within one year in the United States. However, in some foreign countries disclosure or public use prior to patent application may block the issuance of a foreign patent.

There are five main reasons for making a patent application:

1. As an offensive strategy to protect a company's interests.

2. As a defensive action to keep competitors from encroaching into areas of commercial interest.
3. As a trade strategy, possibly useful as a future bargaining chip.
4. For public relations to attract investors who desire tangible evidence of product potential and corporate performance.

With regard to the NPI mycorrhizal product Nutri-Link, patent application is pending on a process for culturing superior strains but not on the strains themselves because they are select strains and not a new construct. The name Nutri-Link is protected by a copyright. NPI expects to continue to protect the product by quality production methods and product improvement.

MARKET STUDIES

Biotechnology companies must keep the consumer in mind during all phases of product development. The marketing plan must focus on the needs of the users and the benefits that the product will offer. Early in the tracking of mycorrhizae development, numerous marketing questions were raised. Inasmuch as the VA Mycorrhizal product we were planning to produce was not on the market, we needed to know what knowledgeable agricultural producers knew about its potential benefits and what their response would be to an entirely new product. We also needed to know who would buy it and what price levels would make it attractive to the buyers. For some of the potential users, formulation might be very important and we needed to know how they would best be able to use this type of inoculant. A final question had to do with finding a suitable name. A marketing consultant firm was employed to probe these and other questions in California and to a limited extent, in Florida. On the basis of our experience and advice from the agricultural consultant firm, we decided to concentrate on the horticultural-nursery industry because many of the plants they produce are grown in nursery beds of sterilized soil or in containers filled with semi-sterile rooting mix.

PRODUCT AWARENESS

The market survey indicated a relatively high awareness of mycorrhizae and the biological activity it supports. University and extension coursework in biology and agriculture for many of the respondents had included concepts in symbiotic soil microbiology, thus preparing them as potential clients to understand the soil-root interactions that form the basis for using a quality mycorrhizal inoculant. Some growers were aware that differences exist among mycorrhizal species and strains and that plants may react differently to the variations in the mycorrhizal populations.

RECEPTIVITY TO USE, PURCHASE

To determine the general receptivity to use a mycorrhizal inoculant, a broad spectrum of citrus, fruit tree, grape, ornamental shrub, and strawberry producers were questioned. These producers have in common the production of plants that are sold as transplants. The plantlets or seedlings are grown on a fumigated soil or potting mix and must have vigorous root systems to be successful transplants. Eighty-three percent of the growers indicated they would try the product under the right conditions. If it was priced reasonably, or free, they would test the mycorrhizal inoculant readily. In general, the survey results indicated an openness to use the product and a belief in the promise of expected benefits.

BENEFIT APPEALS

Based on laboratory, greenhouse, and field studies at NPI and in the literature, we could show that the benefits of mycorrhizal inoculation include increased uptake of phosphorus, water, and some trace minerals. In addition, inoculated plants have been shown to have fewer root-related diseases and to be more vigorous in overall growth. When considered in relation to production costs and the need for a superior product, the respondents to the survey rated promoting a more uniform crop, enhancing resistance to common root diseases and increasing growth in the nursery as the most appealing benefits. On the basis of this finding, our earlier idea that mycorrhizae's most important commercial benefits of reduced phosphorus and water requirement were important to the nursery target group had to be revised in favor of featuring plant quality and performance in the marketing strategy.

PRICE CONSIDERATIONS

Developing a reasonable price for a completely new product must be based on perception of value to the potential users. This must be considered along with the more practical aspect of obtaining a return on investment after recovering the costs of production, packaging, marketing, technical services, and delivery. The major finding was that prospective clients would pay a price for the product in terms of the increased value they could obtain for their plants if they used the inoculant. Growers selling a product with a high unit price such as tree transplants would pay more for the inoculant than those selling a product with a low unit price such as strawberry. Within categories of growers, the lower the projected price for the inoculant, the higher the proportion of growers who would buy it. Interestingly, the more creditable the nursery manager perceived the inoculant to be, the higher the price he claimed he would pay.

RECOMMENDED DELIVERY SYSTEMS

Wherever possible, a product formulation needs to be developed in the most convenient form to fit into the consumer's production methods without causing any extra steps that would increase costs. In response to a multiple-choice question to growers as to how they would prefer to apply the mycorrhizal inoculant, a variety of methods were suggested. Growers of deciduous trees and ornamentals would prefer to add the inoculant to potting mixes. Growers of grapes and strawberries would prefer to mix the inoculant with water to form a root dip. Citrus producers suggested developing a method to add the inoculant to irrigation water. Taking these suggestions into account, both granular and liquid formulations were developed. Each formulation has a high density count of 1000 spores per milliliter or gram.

REACTION TO AN UNKNOWN BIOTECHNOLOGY COMPANY

A general problem that may face relatively new agricultural biotechnology companies is their lack of visibility or reputation. Not only did NPI have concerns that consumer acceptance of a new inoculant product with no previous history would be difficult but acceptance of the company itself might pose a problem because in the target marketing area it is an unknown with a strange name. Interestingly, several of the respondents expressed confidence in a small biotechnology company because they expected it would be more innovative and the product would be supported by high technology research. Most respondents declared that the "unknown company" concern would not be a factor in their decision to buy the product.

PRODUCT NAME

Choosing a suitable name is important in creating a favorable reception for a product as well as introducing its qualities. Reaction to a list of possible product names offered to respondents on a rotational basis elicited a variety of responses ranging from, "that's particularly unappealing," to "that's a name a person can relate to." Some suggested a strategy of incorporating the mycorrhizal idea in the name. From a list of ten names that included Dynagrow, Inoc, Assist, Nutri-Link, Mycron, Paydirt, Symbion, Upstart, Provam, and Booster, the name Nutri-Link was selected. Some names in the list had been used previously in other contexts and were not available. Subsequent development of promotional literature and contacts with growers indicated a very high acceptance of the new name.

PILOT DEVELOPMENT

Making the transition from research to scale-up production requires many innovations. Immediately following success in producing cultures of superior strains of VA Mycorrhizae and characterizing crop plant responses to these strains, plans, and schedules for a pilot production unit were established. Three important elements of the planning process were: building a stage-one scale facility, establishing criteria for quality control, and developing strategies for test marketing.

SCALE-UP PROBLEMS

The transition from laboratory research to volume production requires many adaptations and extrapolations. To achieve the level of product necessary to meet the marketing goal of 2000 liters planned for the initial year, NPI designated a sterile (clean) greenhouse in which to grow the required amount of product. Whereas research had been accomplished on an individual container basis involving growth of select mycorrhizal strains on the roots of host plants in a pasteurized potting mix, scale-up operations required many large pots on a batch basis repeating the operation many times. Planning for scaled-up operations included determining the time for each growing cycle, having sufficient supplies on hand as needed, developing criteria for cold storage of the product while waiting for subsequent batches to be produced, development and testing of an optimum medium for carrying the spores, and packaging to protect the product as well as giving clear and useful information for its use.

Based on the experience received in the primary scale-up operations, many lessons and ideas are being gained that will be useful in the next phase, major-scale commercial production. A more effective way of spore production and separation is being developed for which a patent application has been filed. Continuous-flow spore production is the goal of this intensification phase.

QUALITY CONTROL

Maintaining consistently high quality is important for any product. However, for mycorrhizae, quality control is among the highest priorities because of the adverse consequences any contamination would cause. The quality Nutri-Link product starts with genetically uniform spores produced in clean culture. Every phase of production operations and packaging is being monitored for uniformity and product quality. Looking to the future, new research with a xenic culture methods using living roots as symbiotic hosts in tissue culture will provide a basis for mass-producing pure cultures of superior strains in an industrial-scale process.

TEST MARKETING

To obtain further information on the performance of the mycorrhizal inoculant under actual field conditions, NPI employed a technical sales representative to market the product in California to four distinct client types. The two product formulations were introduced to nursery growers of citrus, grapes, deciduous fruit trees, ornamentals, and strawberries. The sales arrangements with growers involved test applications to these crops that are sold as transplants. The technical sales representative maintains a close contact with the clients to monitor the progress of the treated plants and to assess the expected improvement in plant quality under actual production conditions.

PRODUCTS AVAILABLE FROM PLANT TECHNOLOGY

The driving force in the stock market value of biotechnology companies is the promise of products. Investor expectations for products in the early 1980s were beyond reality because of the complexity of the technologies being employed in plant biotechnology research laboratories. Logically, the more complex the technology, the longer the time required to deliver a stream of products (Figure 2). Products immediately available are clones of superior plant biotypes produced by *in vitro* meristem culture. Some examples of mass cloning of flowers, shrubs, and trees for commercial markets were discussed previously. Disease-free plantlets of vegetatively-

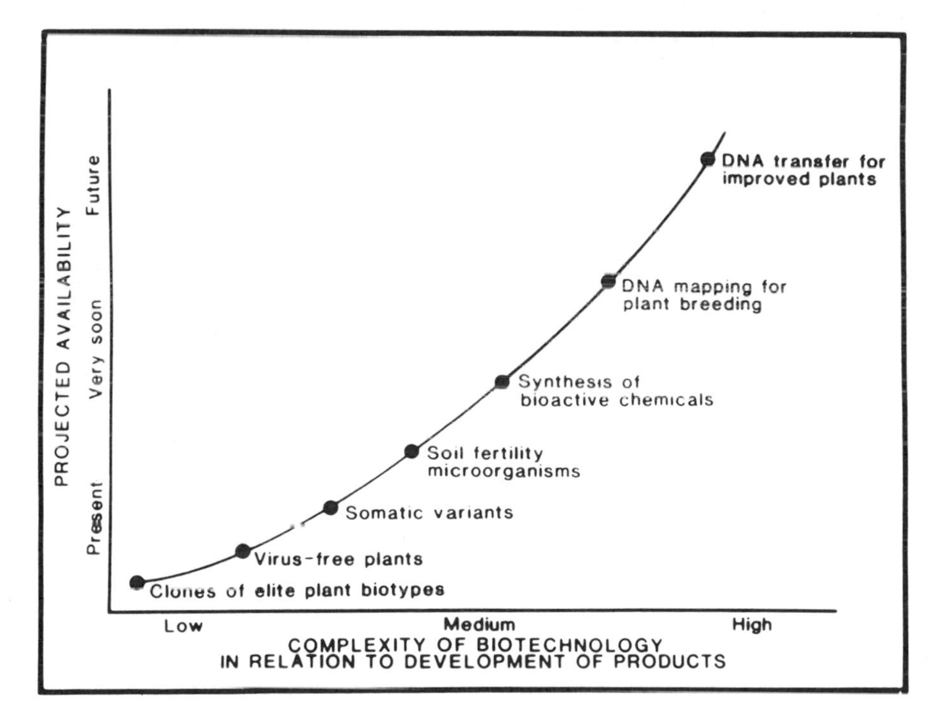

Figure 2. Generalized projection of new products available from plant biotechnology in relation to the complexity of their technologies.

propagated crops such as potato and sugarcane can now be used to produce planting materials from which whole fields can be planted. The result is greater crop yields.

Unique plants selected from somaclonal variants will be the next products from biotechnology (Evans and Sharp, 1986). Reports of somaclonal selections from potato, tomato, chrysanthemum, and alfalfa, among many other species, have been reported at professional meetings. Regeneration of plantlets from fused protoplasts of superior parental genotypes will result in new commercial plant products that are uniquely selected for their combination of traits from diverse genetic sources. Thus, the tissue culture technology holds promise for new products that have been improved genetically without waiting for the more complex technology of recombinant DNA transfer.

Following plant products from tissue culture, we may expect to see improved soil microbiology inoculants. Selected and genetically improved strains of nitrogen-fixing bacteria and growth-promoting selections of mycorrhizae will be followed by recombinant DNA-altered soil organisms capable of retarding the formation of ice crystals, controlling the growth of soil-borne root pests, and promoting various root functions.

A few natural plant chemicals useful for controlling plant pests are now available but in the future many more will be available to improve agricultural efficiency and reduce cultural costs. Because these natural chemicals are specific to the life-cycle of target plant pests, their use will not pose the degree of environmental risk that presently used broad-spectrum pesticide chemicals do. Chemical synthesis of the biologically active portion of the chemical structure of these natural chemicals will allow large-scale production sufficient to make the chemicals available to the agricultural industry at a price it can afford to pay.

Recombinant DNA techniques are probably the most complicated areas of research underway in plant technology and thus commercial products are the farthest in the future. However, near-term use of gene mapping techniques (Nienhuis *et al.*, 1985), using restriction fragment length polymorphisms to follow genetic traits, identify superior genotypes and predict the degree of hybrid vigor of potential matings, hold promise for accelerating plant breeding. Another application of the gene mapping technology is pattern mapping or "fingerprinting" for variety protection. The ultimate and most elegant stream of commercial products from plant biotechnology will be in the production of genetically engineered plants created by the transfer of desired traits into existing commercial plant varieties and species. The prospect of plants adapted to stress, resistant to pests, superior in stature, improved in yield, and higher in natural product quality will be a fulfillment of the hopes and dreams of farmers throughout the ages.

For products of the more complicated technologies to be developed, considerable regulatory cooperation, public understanding, and financial backing still must be forthcoming. The present climate for imposing regulatory controls on biotechnology is too negative. Proposed legislation could restrict new developments far beyond any reasonable necessity and install a regulatory process that could create logjams instead of facilitating progress (McCormick, 1986). However, with confidence that the immediate problems can be solved, new commercial products from the plant biotechnology laboratories, research farms and nurseries will extend the geographic limits of agriculture, increase production efficiency, and add qualities only considered theoretical in the past.

REFERENCES

Aldon, E.F. (1978). Endomycorrhizae enhance shrub growth and survival on mine spoils. pp.174-179. In R.A. Wright (ed.) *The Reclamation of Disturbed Arid Land.* Univ. of New Mexico Press. Albuquerque, N.M.

American Bar Association. (1982). *What is a patent?* ABA Section of Patent, Trademark and Copyright Law. (Brochure) Chicago, IL. p. 16.

Arditti, J. (1977). Clonal propagation of orchids by means of tissue culture—a manual. In Arditti, J. (ed.) *Orchid Biology—Reviews and Perspectives.* I. Cornell Univ. Press. Ithaca, N.Y.

Bonga, J.M. and D.J. Durzan (eds.). (1982). *Tissue Culture in Forestry.* Martinus Nijoff, Dr. W. Junk Publishers. The Hague. p. 420.

Bronson, G. (1985). An idea takes root. *Forbes* 137 (11):134.

Bylinsky, G. (1985). Test tube plants hit paydirt. *Fortune* 112 (5): 50-53.

Call, C.A. and C.M. McKell. (1985). Endomycorrhizae enhance growth of shrub species in processed oil shale and disturbed native soil. *J. Range Manage.* 38 (3) :258-261.

Evans, D.A. and W.R. Sharp. (1986). Applications of somaclonal variation. *Bio/Technology* 4:532-582.

Gerdemann, J.W. (1975). Vesicular-Arbuscular mycorrhizae. In Torrey, J.G. and D.T. Nicholson (eds.), *The Development and Function of Roots.* Academic Press, New York, N.Y. 575-591.

Lawton, K. (1986). Biotechnology, the patent race is on. *Farm Industry News.* March pp. 7-10.

McCormick, D. (1986). Free release regulation still up in the air. *Bio/Technology* 4 (4):273-275.

Menge, J.A. (1980). Mycorrhiza agricultural technologies. Food and renewable resources program. U.S. Congress, Office of Technology Assessment. Washington, D.C.

Murray, J.R. (1986). The first 4 billion is the hardest. *Bio/Technology* 4 (April): 293-296.

Nienhuis, J.T., T. Helentjaris, M. Slocum and A. Schaefer. (1985). Restriction fragment polymorphism mapping of loci associated with insect resistance in tomato. *Agron. Abstr.* p. 133.

Peters, T.J. and R.A. Waterman Jr. (1984). *In Search of Excellence.* Warner Books, Harper and Row. New York, N.Y. p. 360.

Upham, S.K. (1982). Tissue culture, an exciting new concept in potato breeding. *Spudman.* Nov. 14-19.

PART VI
SPECIAL LECTURES

Commercializing Biotechnology Resources: Competition and Cooperation in Global Markets

George Kozmetsky
IC² Institute
The University of Texas at Austin

INTRODUCTION

For the past dozen years or so, biotechnology has been one of the most fascinating scientific explorations since the advent of the transistor and computers. The history of the research and development of semiconductors and the myriad of subsequent exploitations and applications are well known. Just as well known is their commercialization through a cooperative and competitive history—a history that for more than two decades saw the United States attain and maintain its global lead in scientific advances, technology, and manufacturing. Only in the last two to three years have we seen our semiconductor technology and manufacturing leads in very large-scale memory circuits challenged and overtaken by other competitors, especially the Japanese. Furthermore, we are well aware how early on the United States lost its manufacturing competitive edge in semiconductor applications to consumer electronic products. In fact, the consumer product edge is still held by Pacific Rim nations.

We are currently involved in meeting the competitive challenge of Japan and Western Europe in the development of supercomputers and knowledge-based computers for the 1990s and beyond. A challenge encompassing preeminence in scientific and technological advances as well as competition for their manufacturing, marketing, and financing.

Cooperation globally in science and technology since World War II has been an important factor for the United States in exercising a world leadership role. Our nation has provided both means and opportunities for other nations to educate their scientists, engineers, and managers in our institutions of higher learning. There has been a myriad of licensing, cross-licensing, and royalty agreements for the transfer of technology to foreign nations and firms. American firms have initiated a continuing stream of high-technology products to be manufactured overseas.

One of the hallmarks of the U.S. States federal megaproject research programs in the 1980s (SDI and space station) has been to seek and attain cooperative research efforts with our allies. These cooperative actions have served to share research costs, diffuse technology, provide for increasing productivity, and expand real incomes of foreign nations. In addition, they have provided subsequent benefits to our own citizens in advanced technology and lower costs, as well as alternative choices of products.

John A. Young, Chairman of the President's National Commission on Industrial Competitiveness, in his testimony before the U.S. House Subcommittee on Economic Stabilization, stated that "the ability of American industry to compete against foreign competitors, both at home and abroad, is central to our attainment of just about any other goal." He added, "The loss of American leadership in technology markets should have grave consequences for the entire economy since its application is the essential ingredient in the improvement of nearly every industry."

U.S. BIOTECHNOLOGY COMPETITIVE LEADS

The United States scientific explorations in the increasingly competitive fields of biotechnology have been, to this point in time, one of worldwide leadership. There is little question that the United States holds the world's competitive lead in the scientific competition for biotechnology.

Cutting-edge technologies such as those found in biotechnology are the primary drivers for creating newer products and services, as well as emerging industries. Among the industries that will be most directly affected by biotechnology research are:

1. Pharmaceuticals
2. Agriculture
3. Chemical
4. Energy
5. Mining
6. Pollution Control
7. Food Processing
8. Biotechnology R&D Support and Supply
9. Consulting
10. Biotechnology Capital Goods
11. Health Care

There are literally thousands of innovative opportunities for the above eleven wide-ranging industries affected by biotechnology research. Normally, from a business perspective, one looks at biotechnology opportunities in three aspects to understand the impacts of competitive advantages. These opportunities can be viewed as those (1) in an enabling role, (2) in a cost-reduction role, and (3) in a supportive role. Biotechnology can play an enabling role that results in newer products that become a reality because of having special biotechnology capability. Some examples are the discovery of therapeutics and diagnostics, crop improvements, improved animal health, or specialty chemicals. These result in true new products being launched in the marketplace.

In a cost reduction role, biotechnology can assist in replacing existing products in the marketplace. In this case, the competition becomes one of comparative costs and qualities.

The supportive role of biotechnology is related to newer supplies and equipment such as separation equipment, fermentation, reagents, etc. Biotechnology opportunities viewed from the perspectives of enabling, cost reduction, and supportive roles, while important for an individual firm, contribute to building a viable biotechnology industry. Understanding these opportunities is important in formulating public policy issues of competitiveness and cooperation in global markets.

Public policy issues about biotechnology competition and cooperation require a look at maintaining scientific and economic preeminence, developing emerging industries through transfer and diffusion, and encouraging economic growth and job creation. In the past, maintaining scientific preeminence has been a major focus for federal government policies. Federal R&D support, especially through wise investments in basic biological research by the National Institutes of Health and the National Science Foundation, have been instrumental in developing and maintaining the necessary scientific knowledge base that can be used to develop the U.S. biotechnology core resources base. The U.S. Office of Technology Assessment has gone as far as to say that continued declines in research support will threaten the source of innovation for biotechnology, as well as other fields.

The decline in federal support emphasizes the need to know the current state of biotechnology industrial support. As the lead article in the April, 1986 issue of *Bio/Technology* stated, "The First $4 Billion is the Hardest." While NIH, NSF, and the Department of Agriculture funding for biotechnology has been substantially over $4 billion per annum between 1983 and 1986, the total cumulative private support for the private sector biotechnology industries through 1985 has been slightly over $4 billion. Over 63 percent of this fund, or $2.5 billion, have been invested in six companies, one private agricultural consortium, and three research universities. Approximately 60 percent has been invested in the application area of therapeutics, 13 percent in diagnostics, 19 percent in agriculture, and 9 percent for specialty chemicals, research lab supply, and other purposes.

The U.S. biotechnology industries are in their infancy. The current levels of federal funding are stabilizing. Funding excellent new projects is difficult. The majority of the funding has been provided by NIH. The federal funds have been primarily for basic research and development. Currently, negligible federal funds are available for commercialization. Consequently, university research necessary to carry on the next stages for developing viable testing and for promoting intellectual property developments necessary for potential commercialization are not underway. Agricultural biotechnology research is in about the same state. It is funded at least two orders of magnitude lower than medical biotechnology research.

The developments and applications for commercialization by private sector companies are currently being held up for necessary regulatory controls. Our embryonic biotechnology industry, consisting of some 200 companies, seems to be caught up in research and development, which stretched out over time, is increasing investment costs for innovative applications that give us a competitive edge for manufacturing and global marketing. Moreover, a number of American biotechnology companies are taking advantage of other nation's incentives to build their manufacturing facilities overseas in Scotland, England, Ireland, and Holland. These countries not only provide standard incentives of lower tax and free training of work forces, but some countries, like Holland, have provided $400 million in venture capital funds to finance the manufacturing plants. They have also strengthened their own research institutions, including higher education, to support the new biotechnology manufacturers.

Current export regulations and FDA regulations related to maintaining manufacturing in the United States with earlier marketing and distribution overseas are being discussed by appropriate congressional subcommittees. Meanwhile, we are facing much earlier offshore manufacturing related to biotechnology than ever before. In some respects, much like the consumer electronics in the early days of semiconductor advances.

Biotechnology manufacturing offshore is one way of reducing cost and investments by reducing the time involved from basic research to marketing introduction. During the 1960s, the time frame from high-technology development to production used to be on the order of four to six years. However, in the 1970s and 1980s, that has increased to ten to twelve years. The resulting increased investments and subsequent costs have been instrumental in adversely affecting U.S. manufacturers' global competitive position.

It is appropriate to ask, "How have we gotten into this position? How do we go about solving these problems?"

PARADIGMS FOR COMMERCIALIZATION

For much of the period from the 1950s to the 1980s, the United States generally assumed that scientific research would, in one way or another, transfer naturally into developments or technologies and subsequently be commercialized. For most of this period, not enough attention was paid to how science was transformed into technology which was subsequently transferred for specific commercialization purposes and then diffused throughout all industries, regionally as well as among our allies. The traditional paradigm has been that basic research innovations supported by government would be utilized by the private sector for applied research and developments and that their manufacture would automatically follow. Diffusion to other uses would occur when R&D results were both economical and better understood by the private sector in general. The utilization of technology as a resource was perceived as an individual institution's responsibility—university and private firms. It was expected that all regions of the United States would in time enjoy the benefits of this paradigm in which new innovations from research were followed naturally by timely developments, commercialization, and diffusion.

Because of the demands from global competition and cooperation, a new paradigm for the commercialization of biotechnology needs to be implemented. This paradigm involves newer institutional developments. These newer institutional developments complement and extend more traditional institutional relationships and rules. The newer institutional developments are providing a set of coherent relationships among key institutions focusing on national pride, U.S. competitiveness, academic excellence, economic growth, and technological diversification.

The drivers for these newer institutional developments are:

1. A desire to foster more basic research.
2. Shortages of adequately trained scientists and engineers.
3. Difficulty in keeping up to date with developments.
4. Gaps in new technology transfer especially when it requires pulling together pure research from different disciplines.

5. A need to fill the gaps for diffusion of technology for developing timely, useful commercial products and services by individual companies.
6. Increasing international competition.
7. A determination to diffuse R&D activities across wider geographic areas.

These drivers are taking place within the context of a hypercompetitive environment. In today's hypercompetitive world, there is a need in each community for leadership that reassesses, if not transforms, economic growth. "Hypercompetitive" is a word that connotes a new economy is emerging in our American society as a result of intense domestic, as well as fierce global, competition. The competition is between states, cities, universities and colleges, industries, all sizes of business, and between highly industrialized and emerging foreign nations.

FORMS OF INSTITUTIONAL DEVELOPMENT

At least eight forms of institutional development for comprehensive commercialization of biotechnology should be considered. They are:

- industrial R&D joint ventures and consortia
- academic/business collaboration
- government/university/industry collaboration
- incubators
- university/industry research and engineering centers of excellence
- small business innovation research programs
- state venture capital funds
- commercialization of university intellectual property

These institutional developments demonstrate the willingness of individuals and institutions to take and share risks in commercializing science and technology. This process provides a means to make and secure the future of the participating institutions. Furthermore, the process establishes a newer infrastructure to support entrepreneurship, encourage innovation, and accelerate technology transfer and diffusion.

These eight institutional developments for economic growth and diversification can be categorized as: (1) encouraging emerging industries; (2) providing seed capital for early and start-up entrepreneurial endeavors; and (3) assuring U.S. economic preeminence.

1. **To develop emerging industries.** Institutional relationships involved here are academic and industrial collaborations, and industrial R&D consortia. Because the basic research is carried out in the universities and colleges, getting collaborative efforts between academia and industry can accelerate the commercialization of basic research into emerging industries. Industrial joint ventures and consortia are still expanding into newer areas that encompass both high-technology and basic industries.

2. **To create seed capital for small and take-off companies.** Some forms of institutional developments such as incubators, SBIR programs and state venture capital funds are providing seed capital for small and take-off companies. They are also pushing regional and local economic diversification through entrepreneurial activities.

3. **To provide for U.S. economic preeminence.** A number of institutional developments are seeking to ensure U.S. economic preeminence. These focus on the creation of NSF centers of research and engineering excellence, government/business/university collaborative arrangements in technological areas, industrial R&D joint ventures and consortia, and NSF's sponsorship of Industry/University Cooperative Research Centers (IUCR). These are intended to provide a broad-based research program that is too large for any one company to undertake alone.

BIOTECHNOLOGY—NEEDS AND GOALS

Perhaps the most counter-intuitive aspect of the commercialization of biotechnology, to me, is the recognition that we need people who are multi- and inter-disciplinary in nature, who can creatively and innovatively transfer technology, and who can successfully link security, profit, and pride in a hypercompetitive domestic/international environment. We do not have enough such talent. Nor are we educating them or deliberately developing this newer talent through successful experiential learning.

Talent needs to be supported in innovative ways. Through innovative forms of institutional developments, it is possible to create the infrastructure to accelerate technology transfer. I am sure that, as the program evolves, the biotechnology field will utilize many of the institutional developments for technology diffusion and commercialization that can develop emerging industries and create small and take-off companies. It can provide the impetus for the education and training of the new core of creative and innovative talents.

Developing biotechnology as an emerging domestic and global industry can be an important factor in maintaining and enhancing U.S. global competitive and cooperative posture. In this sense, we must accomplish the following goals:

1. Assure the continuance of our national preeminence in the necessary core sciences of engineering and biotechnology.
2. Develop and maintain a favorable balance of trade for export market for biotechnology products, services, supplies, and capital equipment.
3. Maintain a significant world market share of the biotechnology markets.
4. Develop the necessary abilities to maintain superior productivity in emerging biotechnology manufacturing.
5. Encourage developments in biotechnology that allow Americans to earn a higher standard of living in an interdependent and increasingly competitive world economy.

DEVELOPING MANUFACTURING CENTERS

It is generally conceded that our biotechnology innovations are our greatest competitive advantage. Even the Japanese admit that we are at least four years ahead of them. A recent *Science* editorial by Daniel E. Koshland, Jr., accurately captured the current state of biotechnology:

> Today's recombinant DNA biotechnology brought with it the usual concomitants of basic research, applied research, and happy accidents. Some individuals worry that science has blurred the lines of demarcation between universities and industry. Their concern is probably justified. Others worry that many biotechnology companies are in financially shaky condition. They are accurate observers. Still others are euphoric about biotechnology: they await a giant flowering after a rather anemic beginning. Their expectations may also be justified. For the accomplishments of the still infant biotechnology industry, its promises of new procedures and new products, to say nothing of its ever more efficient production of goods, are awesome.

Our biotechnology industry in 1986 consists of more than 200 firms. A large number of universities are conducting research and are being supported by NIH, the Department of Agriculture, and NSF. State governments are including in their budgets allocations to improve the faculty and facilities for science programs. Some states have programs that include regional high-technology centers that stress research and development in biotechnology. Others have established research parks as well as incubator facilities for start-up companies. Some states have provided financial assistance to support basic and applied research within their higher educational institutions.

The major issue facing the United States in the area of biotechnology commercialization is the development of manufacturing centers. More specifically, up to this point, we have concentrated on the development of world class innovation centers. By that I mean centers of research and development excellence. Two examples of such areas are Route 128 and Silicon Valley. Subsequently, other nations have licensed our technology and then proceeded to develop manufacturing centers. Through these centers they have, over time, made significant advances in product designs, cost reductions, and subsequent taking over of increasingly larger shares of world-wide markets. In other words, they have focused on the commercialization of many of the technological advances which have emerged from our innovation centers.

How do we go about developing and ensuring U.S. domestic and global market leadership in the commercialization of biotechnology? First, we must recognize that we have been developing a number of critical infrastructure components. Our research and development is world class. We need to assure ourselves that we maintain this competitive edge especially in those market areas necessary to attain America's security and goals in an environment that stresses cooperation. This means that over time other nations will develop their own appropriate centers of excellence in biotechnology which in turn will contribute to their share of world-wide markets. Each nation can have its own technological resources and commercial competitive capabilities. Cooperation takes place for basic science and other developments, including substantive theory and methodologies. In other words, Nobel prizes can be shared as well as the basic knowledge. In my opinion, it is difficult, if not impossible, for any one nation to maintain scientific superiority in all areas of biotechnology.

Consequently, we need to reevaluate our own domestic competitiveness for scientific superiority. There is little reason to have each of our 50 states and hundreds of research institutions competing against each other. Especially if they ignore the needs and subsequent markets within their own states. To mitigate such hyperactive competition in the developments of core sciences and methodologies, it will require newer collaborative efforts, institutional relationships, and advanced developments of telecommunications. There is enough technology to network our research scientists and engineers in our federal government laboratories, industry research laboratories, universities and colleges, associated federally funded R&D centers, and other nonprofit research centers to better disseminate the results of the core knowledge and know-how. More importantly, we must find ways to reduce the time and effort to transfer and diffuse our biotechnologies. Most importantly, we must use the newer telecommunication technologies and supercomputer abilities to diffuse the biotechnology knowledge and know-how to be utilized more expeditiously by medium and small firms.

Equally important to the success of commercializing U.S. biotechnologies is the improvement of communicating their impacts and benefits. To wait too long to develop the necessary policies and regulations at the federal, state, and local levels results in confusion, unnecessary delays, and increased investments and costs. It also threatens our competitive edge for world-wide markets. The consequences on economic growth and job creation may be more detrimental than return on investment or short-term profits to the private sector.

In time, we must learn to expand the communication network for maintaining our competitive edge in the U.S. biotechnology industries. Each firm needs to have access to timely and accurate information ranging from core scientific and methodological advances to domestic and global market data and needs, to other information for enhancing its abilities to compete and cooperate.

Today's major U.S. public policy issue is how to most effectively develop and improve our manufacturing abilities in biotechnology. Successfully scaling up R&D applications is a critical barrier. To date, there are not only technical limitations but also regulatory ones for manufacturing facilities.

THE JAPANESE APPROACH

Such steps were taken by the Japanese in 1982. They increased their recombinant DNA mass cultures from 20 liter capacity to 300 to 500 liter tanks. They decontrolled microorganism containment standards from P3 (decompression of experimental room's internal atmosphere to lower level than outside air) to P2 (airtight and constantly kept in a state of more outflow than inflow of air through no microorganism filter). They also eased experimental guidelines by amending rules for mass culture of microorganisms as long as the equipment met a prescribed standard in order to give a go ahead to industrial applications of DNA technology. Prior to this, government agents checked safety factors. They also approved plant host experiments in order to allow for the diffusion of research from pharmaceuticals to foods to agriculture.

In the typical Japanese strategic approach, they established two new institutes for fuller utilization of biotechnology for Japan. These institutes are the National Institute of Agrobiological Resources and the National Institute of Agrobiological Sciences. They were directed to establish the knowledge and know-how to utilize biotechnology for the 21st century. They are newer institutional developments supported by the government and industry. Some of their work is to develop plant improvements by cell fusion, efficient production of seeds by tissue culture, and bioreactor systems for food production. They also created an international seed center for Japan and Southeast Asia.

ISSUES IN MANUFACTURING

The manufacturing of U.S. innovative applications is currently being discussed as follows:

Perils of Integration—When can and will start-up companies become fully integrated pharmaceutical houses? Some believe that one can build a drug company on one successful drug. Licensing from a start-up enterprise often means giving up a new company's future. Others believe small companies cannot afford to finance and survive for the time it takes to commercialize and market after clinical trials, needs for unusual observations, and additional testing and arduous approval processes.

Integrating Forward—This involves start-up companies linking with giant companies—domestic and foreign—already in the manufacturing and marketing.

Large versus Small—It is difficult to make a pharmaceutical company out of good science alone. Small companies need to plow back their earnings from revenues into R&D in order to survive in their niche markets. They can build up their equity base and market value by building the technology base. They can have better R&D capability than in some integrated bureaucratic large companies. Of course, there will be more mergers and acquisitions over time as a means to realize values developed and earned.

Viability of R&D Biotechnology Small Firms—Since 1974, small businesses (under 50 employees) have provided over 64 percent of the employment in the United States. Will this continue in the biotechnology industries? Some experts believe that the number of smaller and medium-sized biotech companies will shake out from the current 200 to about 50 or so that can serve 2,000 or so global pharmaceutical, chemical, and agricultural companies. These companies may become much like the specialization in contract research.

The United States has a commanding lead in biotechnology science and engineering. We have utilized our venture capital industry as well as limited R&D partnerships to develop a critical mass of innovative R&D firms. We have yet to evolve into the manufacturing phase of biotechnology. There are hundreds of therapeutic and diagnostic products in the regulatory approval process. There are no manufacturing centers for the newer processes that will be required for world-wide competition. This is true of most other high technology developments in the United States.

Manufacturing is a national issue. We have seen our imports in five of seven high-technology areas exceed our exports. In addition, U.S. firms are currently placing more high-technology manufacturing directly or by joint ventures overseas. Meanwhile, Japanese firms are establishing manufacturing as well as their own suppliers in the United States. There are over 34 Japanese firms currently involved in establishing manufacturing entities in the United States. MITI is in disagreement with these firm policies while the Japanese Economic Agency approves. They obviously have their own internal disagreements. European and Canadian companies are, and have been, actively involved in acquiring American companies. Between 1983 and 1985, over 450 companies were acquired by foreign companies. Only 29 companies were acquired by the Japanese. The United Kingdom acquired over 140 companies, and our neighbor, Canada, acquired over 100 companies.

We have not concentrated or focused our resources, scientific and technological abilities, and academic expertise on developing U.S. leadership in manufacturing. We have relied for some time on the invisible hand to develop innovative applications that resulted from our federal core research advances. We have had very little of our federal agencies' support applied to biotechnology and commercialization. State governments, particularly since 1980, have developed newer institutional developments aimed at economic growth, diversification, and job creation. The successful creation of jobs in the past decade points to their successes. However, neither the federal nor state governments, along with small and medium-sized companies, focused on improving and accelerating U.S. manufacturing capabilities and competitive posture.

Fortunately, biotechnology is still in its developmental stages. This permits us the opportunity to take advantage of our current core science and engineering leadership and develop the required American manufacturing competitive posture while cooperating in the diffusion of our advances.

In the final analysis, America's strength has always been its ability to be scientifically creative, technologically adept, managerially innovative, and entrepreneurially daring. The current state of biotechnology demonstrates the first two abilities. It has the attributes of being scientifically creative and technologically adept. Successful commercialization now depends on being managerially innovative and entrepreneurially daring.

ACKNOWLEDGMENTS

There are a number of friends and colleagues whose inputs and critiques have been especially insightful and helpful in the development of this paper.

I would like to express my appreciation to Drs. Hans Mark (The University of Texas System), Gerhard J. Fonken (The University of Texas at Austin), John Prentice Howe, III, (The University of Texas Health Science Center at San Antonio), Charles A. LeMaistre (The University of Texas System Cancer Center), Charles C. Sprague (The University of Texas Health Science Center at Dallas), William C. Levin (The University of Texas Medical Branch at Galveston), Roger J. Bulger (The University of Texas Health Science Center at Houston), and Jay Stein and Norman Talal (The University of Texas Health Science Center at San Antonio).

From the IC² Institute of The University of Texas at Austin, I would like to acknowledge the constructive suggestions of Eugene B. Konecci, Ray W. Smilor, and Tom J. Mabry. I also wish to thank Linda Teague for her assistance in preparing this manuscript and Patricia Roe and Ophelia Mallari for their research assistance and other support.

Tomorrow's Science Policy and Biotechnology

Orville G. Bentley
Assistant Secretary for Science and Education
U.S. Department of Agriculture

INTRODUCTION

It's a pleasure to be here with you today to speak to such a distinguished group. Dr. Price and Dr. Tefertiller have done a terrific job at putting together a comprehensive program.

It takes a lot of hard work and money to make events like this happen. We're all grateful for the generous sponsorship of the University of Florida, The University of Texas at Austin, and the RGK Foundation.

Furthermore, in dealing with a subject as dynamic as biotechnology, it is important to have the interest and support of people like Senator Chiles. His willingness to spend a day here listening to recommendations is evidence that he fully appreciates the magnitude of the issues and questions involved.

The title for my talk with you today is "Tomorrow's Science Policy and Biotechnology." I'll tell you right off, I feel pretty humble in a crowd like this making my predictions for "tomorrow." They may sound a bit conservative to some of you. Just remember, what I say represents merely one four-billionth of the world's opinion. If I go out on a limb, I just hope nobody brought a saw.

Predicting the future can be a hazardous business—especially if anyone remembers what you say—and somebody always will.

For example, people still remember that Lord Kelvin, the eminent nineteenth-century physicist, went on record as believing that "X-rays will prove to be a hoax...aircraft flight is impossible...and radio has no future."

It proves that scientists are certainly not immune to erroneous predictions. Even Albert Einstein got caught. In 1932, he wrote "There is not the slightest indication that (nuclear) energy will ever be obtainable."

Since it's so risky looking into the future, it makes sense to ask why we keep on doing it.

First of all, we do it partly to reassure ourselves—to create the illusion that in this fast-paced world where life rushes past us at such a terrific speed, we really do have some idea of where we're going and what we'll find when we get there.

The second reason—and decidedly the more optimistic one—is that it enables some special few who are possessed of insight and wisdom to share that vision with the rest of us.

Al Wood was that special kind of person. He wasn't afraid of the word "tomorrow" as he preached the promise of biotechnology. He helped us all to understand the scope and the pervasiveness of modern molecular biology in the field of agriculture.

The hand and the hard work of Al Wood are in evidence throughout this event. He was a devoted scientist and a true friend. His presence is greatly missed.

On May 23, *The Wall Street Journal* quoted a scientist who said: "Genetic engineering is the second big bang in biology." We didn't have anything to do with the first big bang, but look at us now. We have learned how to manipulate and move the genetic elements around. The possibilities are tremendous.

And they are more than possibilities. They are realities. You heard about some of them this morning from outstanding leaders in the field—and you'll hear about more exciting developments as the day goes on.

I can't add a great deal to this erudite discussion of the latest advances of science and technology, but I would like to talk about some of the other advances that need to take place at the same time.

The future of biotechnology depends on many interrelated factors. I want to discuss three of them with you today: (1) results, (2) regulation, and (3) responsibility.

RESULTS

Biotechnology has been around a long time. The June *Science Digest* list of "Milestones in Biotechnology" starts with before 6000 B.C. when yeasts were used to make wine and beer—then it goes to 4000 B.C. and leavened bread—and then a tremendous jump to Gregor Mendel in 1865. Once into the 1970s, things really start popping with crucial discoveries right and left.

That list sets out what has been accomplished in biotechnology in the past. The important question at this point is, "What are we accomplishing now?"

Many of our institutions continue to act as if tomorrow were going to be very much like today. Their way of predicting the future is simply to extend the status quo.

I prefer the outlook of John Naisbitt, author of *Megatrends*. He says that "things aren't going to get better. They're going to get different." And I would add that they're going to get different very fast. We must prepare for the changes.

Molecular biology is a powerful new scientific tool, but it only works if you use it. Appropriate technologies should be widely available and interdisciplinary in scope. That belief was emphasized in a document which we all associate with Al Wood, *Emerging Biotechnologies in Agriculture: Issues and Policies*, Progress Report II, November 1983, Division of Agriculture, Committee on Biotechnology, National Association of State Universities and Land-Grant Colleges.

The 1983 edition reads, "Maximum success in the application of the new research technologies to the enhancement of agricultural productivity will require an integration of the basic sciences...with the disciplines of traditional agriculture science..."

"This (integration) will require different kinds and degrees of program coordination depending upon the hierarchical structure of the academic institution. By their very nature and structure, the land-grant universities provide an ideal setting for the application of advances in biotechnological research to agriculture."

Let's look at that research effort. What are USDA and the State Agricultural Experiment Stations doing? Up front, let me assure you that we have a strong commitment to fundamental science and to the technological application of biotechnology to our agriculture, forestry, and food and fiber systems.

And we are backing up that commitment. About $26.3 million of the ARS budget is for projects that utilize biotechnology in the research protocol. And under the competitive grants programs of the Cooperative State Research Service, $28.6 million was allocated in the 1986 fiscal year.

The states themselves have an investment in biotechnological research equivalent to 373 scientist years, 583 students, and 650 staff members. I'm sure you can appreciate the sizeable commitment of funds which that represents.

Spending money is necessary and important, but along with advances in biotechnological research, we must make advances in the transition between the laboratory and the marketplace. In order to derive the full benefit of biotechnology, we must perfect the art of turning scientific discoveries into business opportunities.

As John Diebold points out in *Issues in Science and Technology* (Spring 1986), there is no lack of examples of U.S. achievements in getting science to market. From jet engines to computers, we have excelled in the process of applying scientific insights to commercially profitable ventures and thus to widespread practical use. American agriculture is an outstanding example of that ability.

While progress in biotechnology can create opportunities for us, it does not by itself create economic growth. It, therefore, becomes important for government to have a strong commitment to the role of science in the national well-being and to the entrepreneurial spirit—the freedom and initiative to explore new markets.

Ian Ross, President of AT&T, stated that "Technological innovation...is seen as a major force for improving the nation's productivity, industrial competitiveness, and economic growth...Most economists concur that 50 to 60 percent of our economic growth can be attributed to technological innovation."

He was speaking of technology in the broader sense, but the idea holds for biotechnology as well. According to the *New York Times*, some analysts predict that biotechnology product sales in agriculture alone could top $5 billion by the end of the century.

American agriculture in particular is looking to biotechnology for lower unit production costs and enhanced quality of products to maintain U.S. competitiveness in international trade.

I want to underscore the critical need we have as a nation to maintain a world leadership position in molecular biology. Our international colleagues expect us to continue to move forward and to do our share in this critical endeavor.

REGULATION

The point that's bound to come up next is that these necessary results do not occur in a vacuum. Success in biotechnology can depend as much on government regulation and involvement as on the quality of a product.

Industries cannot afford to wait three or four years in order to test their products. As the *New York Times* points out, it's unwise, even tragic, to spend time pushing papers and paying lawyers while baby pigs die of pseudorabies or the Florida citrus crop freezes on the trees.

Government is naturally a supportive partner of the fledgling biotechnology industry and the contributions it can make to the quality of life. At the same time, we must, understandably, address public concerns and perceptions.

The United States has a broad base of laws and regulations designed to protect the environment and promote public safety. These laws serve a useful purpose by influencing the process of applying research results and technological developments to actuality.

We believe that the new government framework for coordinated regulation of biotechnology—which we expect President Reagan will soon sign—recognizes both the reason for this protection and the need to base such decisions on sound science.

The new guidelines will relieve minimally risky products of the harassment that can chill their progress. They will permit carefully controlled evaluation experiments to move ahead. They should help loosen the death grip that federal court injunctions and local prohibitions have had on the orderly development of biotechnology research for agriculture.

USDA played an active and involved role in the formulation of these guidelines. This administration has invested a tremendous amount of effort into creating a unified framework under which federal agencies can work together to regulate biotechnology and its products.

USDA believes that the agriculture and forestry products developed by biotechnology will not differ fundamentally from conventionally produced products. While we feel that the existing regulatory framework is adequate, we have pledged to amend our regulations if developments indicate a need to do so.

We want to encourage promising developments in agricultural biotechnology, in both USDA-funded research projects and in commercial product development, while protecting human health and the environment.

It is our hope that the newly-authorized Office of Agricultural Biotechnology will do just that. At the direction of Secretary Lyng, the Assistant Secretary for Science and Education will establish the OAB and it will be responsible for implementing and coordinating the department's biotechnology policies and procedures.

This includes the conduct of laboratory and field research, experimentation on biotechnological products prior to their commercialization, and all matters of oversight of biotechnology in agriculture.

RESPONSIBILITIES

Carl Sandburg wrote that "Nothing happens unless first a dream." Biotechnology has been a dream for a long time.

Now comes the hard part. As William Butler Yeats wrote, "In dreams begins responsibility."

Though neither poet was writing about biotechnology, their statements are universal. While the ability to manipulate genes within their cells has the potential to provide mankind with incalculable benefits, it brings heavy responsibilities as well.

We in the technological community are faced with twin obligations—first, to turn opportunities into realities—and second, to be responsible scientific watchdogs.

Only in this way can we freely begin the developmental process of not only judging the efficacy of new innovations but of evaluating the possibility of environmental risks.

The stakes are high. Public opinion surveys over a number of years by the National Science Foundation and the National Science Board indicate a steady increase in public awareness of science and technology issues.

As National Science Foundation Director, Erich Bloch correctly observes in *Issues in Science and Technology* (Winter 1986), "Public demands on science and technology will intensify because of (our) increasing impact on the nation's economy and security and (our) influence on the public's work, health, and leisure."

In her article, "Genetics and the Law: A Scientist's View," Maxine Singer makes a strong plea for science to take action: "Today, in the United States, the threat which results from the scientific ignorance of policy-makers must be faced. Genetic engineering is only one area of science in which this ignorance can lead to disastrous results. Education can end this ignorance...Is it not time we began?"

It is clearly up to us to make sure that we continue to throw light on the subject of science—and that we work to reduce our isolation from public attitudes and from political debates about science and technology, and their consequences.

As we push toward the frontiers of biotechnology, we must work to bring the public along with us. If not, we will be like the frontier scout who rides out far ahead—and suddenly looks back to discover that he's lost the whole darn wagon train.

In addition, we must make every effort to be fully aware of the possible social and economic implications of biotechnology. Technology-driven changes have the potential to restructure American agriculture and to sway the social and economic viability of rural institutions.

And, perhaps closer to home, what changes will biotechnology bring about in our research and education program? We must work out a thoughtful plan which will enable us to enunciate a coherent policy for the future. We must preserve a long-term perspective.

CONCLUSION

When I see the wonders and the opportunities and the challenges in science that lie before us today, I experience a strong optimism about the future of science. It's enough to make me wish I were a graduate student again—unless, of course, I had to go back to washing my own test tubes.

When Walt Disney created Disneyland, one of the four lands he imagined was Tomorrowland. To him, Tomorrowland was full of rocket ships and robots and travel to the planets. The future was outer space.

You and I are fortunate, however, to be part of another future—the **inner** space of biotechnology—the universe of the molecule—the galaxy of the gene.

For good or ill, not leaving well enough alone is what human nature is all about. As part of mankind's eternal urge to dispel the darkness, we continue to push at the boundaries of science and technology.

Peter Raven, head of the Missouri Botanical Garden, was one of the scientists who initially opposed the work done by Steven Lindow on "ice minus." Now, in an article in the May 26 *Newsweek*, he says that while he is "still concerned," he supports the tests. "Without tests," he admits, "we can't begin to evaluate the wider consequences."

As *Newsweek* concludes, "Given the history of technology, it's possible and perhaps likely that someday, someone will misuse genetic engineering and create an environmental problem."

But I'd like to close with botanist Raven's statement on the future: "To pretend we're living in a pristine forest and say we shouldn't change anything is absolutely absurd. In the end, using biotechnology to raise the agricultural productivity of areas we've decided to cultivate may be the best way to leave other parts of the world unaltered."

I believe that we have within ourselves the strength and the spirit to deal successfully with the challenges and the opportunities of biotechnology—if we approach them with caution, vision, and sound science.

Sustaining American Leadership in Biotechnology

Robert Rabin
Assistant Director for Life Sciences
Office of Science and Technology Policy
The Executive Office of the President

It was at the University of Florida on a warm day one late spring that I first met and was so impressed with Al Wood. His enormous energy and vitality are missed here and on the national scene also. With clarity of purpose and vigor that few match, he channeled his energy to strengthen the scientific base of agriculture. Al would have enjoyed this symposium and being among his colleagues jousting with their various points of view. You who have organized the symposium have done well and your dedication of the proceedings to him touches us.

The phrase, "No stone was left unturned," came to my mind as descriptive of the last two days' discussions. This thought quickly reminded me of the story once told in the *Wall Street Journal*. Volunteers were cleaning sea birds incapacitated by an offshore oil spill. The suggested chemical cure proved as bad as the problem. The headline over the story read: "Solvent leaves no tern unstoned."

A title of this talk was suggested to me along the lines of "How To Sustain American Leadership in Biotechnology." Some answers are glibly and simply stated, but implemented with difficulty. Try this answer for example: Strongly increase funding for research related to biotechnology. Here's another: Minimize confusion attending the regulation of such research and the products derived from it. Or this: Write a law that gives a large tax break to firms which collaborate with university scholars and donate equipment.

One of the problems in sustaining American leadership in biotechnology is defining leadership. There are at least 10 types of stakeholders in biotechnology: the universities, the National Academies and professional societies, the individual firms and their trade associations, the public interest groups, the executive branch of the federal government with its network of public advisory groups, the Congress, state and local governments, the public, the press, and the legal profession. Each type wants to be the leader or to influence the leader; each has invested heavily in ego, jobs, and money; each is convinced that its cause or its concerns are just, even uniquely so.

I could dwell on the fascinating and sometimes frustrating sociology of American biotechnology beyond your post-dinner endurance. I choose, instead, to consider with you a chronological—an historical—bimodal panorama and the polar aspects of the scientific community itself. Remember the result of the 1973 Gordon Conference on Nucleic Acids? The journey started with the letter to *Science*, with stopovers at the National Academy of Science, and the Asilomar Conference in 1975. It ended with establishment of the RAC—the NIH Recombinant DNA Advisory Committee. There followed the Cambridge and Princeton Public Debates, and congressional hearings that seemed to be heading on a fast track to the legislation of research.

Within the community of molecular biologists, disarmament referred far less to strategic arms limitation treaties than to biological containment. The hyperactivity subsided as gradual relaxation of the NIH Guidelines dispelled public fears that monsters could escape. Page one stuff of the early and mid-decade became non-events. The Foundation on Economic Trends wasn't even a gleam in Jeremy Rifkin's eye. We biologists needn't have despaired for lack of attention because the herpes and AIDS viruses were waiting, as were the new microbial equals of the mythological chimeras—organisms destined, indeed designed, for environmental release.

So the first mode of the bimodal panorama expired without legislation, without mishaps. The NIH RAC did a superlative job and was trusted by academicians, industrialists, the Congress, and the public. The guidelines, intended for the safety of the laboratory worker and for containment of the organisms, were adopted almost universally by federal agencies supporting academic research. Industry voluntarily abided by them. As the first mode was ending, the new biotechnology industry was beginning. Fueled by venture capital, Wall Street equity offerings and entrepreneurs, biotechnology again became front-page news, at least in the business section. Those who had debated, even anguished over, the ethical concerns of catalyzing evolution were still heard, but fewer seemed to listen.

The second phase of the bimodal history is upon us. If the recombinant or genetically engineered organisms were mainly safe, couldn't they be designed to carry and express useful traits applicable to agriculture and forestry in the environment? The answer from some clearly was "yes." We could insert exact nucleotide sequences in the genome which would code for the production of proteins. Better yet, we could precisely regulate their expression and activity with flanking sequences whose identity and functions were known in advance. The beauty of these regulatory sequences—these origins of replication, ribosome binding sites, promoters, operators and terminators—is that they don't produce a gene product. They only initiate and modulate nucleic acid synthesis at a specific region. They are the governors of the business.

The changing ethos surprised many. We had accepted containment; deliberate release was now antiethical. Rancor and litigation replaced accord. Molecular biologists and geneticists, who had dominated biotechnology, were aggressively challenged by ecologists.

The arguments centered on the lack of experience and information about the fate and environmental effects of introduced genetically engineered organisms. Widely accepted test protocols to provide data on survival, multiplication, dispersal and effects were not available. The NIH RAC struggled with the problem. Its environment subcommittee recommended a series of "Points to Consider" in conducting research that involved intentional release. But it was clear that the government needed more than NIH to be the spear carrier for its agencies as the new organisms had to be moved out of the lab for testing under natural conditions. Also, NIH could hardly be blamed for reluctance to be in court on issues with which it had little experience and which were allied much more closely with other agencies' research and regulatory functions.

In April 1984, Jay Keyworth, then the President's Science Advisor, called policy officials of departments and agencies to a meeting in the Old Executive Office Building. I attended as the representative of the National Science Foundation. He delivered a plain message: The developing industry doesn't understand the regulatory labyrinth of biotechnology. Make your regs and policies understandable. You will continue to meet as a newly constituted Cabinet Council Working Group on Biotechnology. You will report to the Cabinet Council on Natural Resources and the Environment.

Eight months later on the last day of the year, the Working Group's labor was published in the Federal Register as the Coordinated Framework for the Regulation of Biotechnology. It also described the Biotechnology Science Board, an overarching policy apparatus with quasi-regulatory power. During the following three-and-a-half month public comment period it became evident that this first effort to forge a national policy was flawed.

On other fronts, things were going poorly in the spring and summer of 1985. Member nations of the Organization for Economic Cooperation and Development were attempting to agree on safety considerations for industrial, agricultural, and environmental applications of recombinant organisms. The U.S. team from the federal agencies was split and could not agree. To prevent embarrassment and quell congressional inquiries, the State Department took the lead and gave its proxy to Frank Young, Commissioner of Food and Drugs. Finally, in December, the issues were resolved among the representatives of the member nations. Full acceptance of the report by the OECD Council is assured.

In June of 1985, in Philadelphia, molecular biologists and ecologists heatedly debated the issues of intentional release under the aegis of the American Society for Microbiology. Since this was the first time all sides squared off in large open sessions, the best to be hoped for was clear statements of positions. They were admirably summarized by Ed Adelberg of Yale but, other than agreement to keep talking, the groups seemed far apart on the safety issue.

Several congressional committees were readying new hearings and bills. Congressional leadership under Albert Gore, Jr. had been vacated by his election to the Senate. Mr. Durenberger, his senior in the Senate, and Representatives Dingle and Fuqua in the House were active, but not fully visible. But Mr. Gore took November 14, 1985, as his day for confronting the executive branch. His timing was perfect. EPA had just approved the field test by Advanced Genetic Sciences, Inc., for the ice-minus pseudomonads prepared by the method of Lindow and Panopoulos. The Office of Science and Technology Policy also on that day published in the Federal Register its decision to establish the Biotechnology Science Coordinating Committee (BSCC). Mr. Gore chairs the Environmental and Energy Study Conference. Although it is not a committee or subcommittee, and he was the only Senator on the dais, the room was filled with people and TV equipment. For those of us who had to testify, it was uncomfortably warm and bright from the lights (and also from the Senator). He attacked the new OSTP committee as being little more than a discussion group with no authority.

In December, Mr. Dingle's Subcommittee on Oversight and Investigations held hearings, and Senators Durenberger and Baucus introduced S. 1967, a bill to amend the Toxic Substances Control Act. Murphy's Law struck bio- and engineering technology with vengeance. NASA lost three space vehicles and one crew; the Chernobyl power plant exploded; Advanced Genetic Sciences had its license to continue its tests of ice-minus bacteria suspended; and the USDA was called before congressional committees to explain how Biologics Corporation was permitted to test, and licensed to manufacture, its pseudorabies vaccine. Last March, Mr. Fuqua introduced H.R. 4452, the Biotechnology

Coordination Act of 1986. Since OSTP is directly and strongly affected, and will have to testify tomorrow, I can only indicate that the Administration does not support it. By the end of May, technology everywhere seemed to be on the defensive.

I return to the suggested topic: how to sustain American leadership in biotechnology. In this brief historical tour, we have touched on events that were shaped by those who aspired to leadership and those who attained it, however briefly. In our democracy, consensus is achieved often with difficulty; lately, in biotechnology, it seems to have avoided us because no sustaining leadership is visible. The scientific community has disagreement within its ranks, the Congress appears bent on new regulatory legislation, the industry is impatient with government, and government agencies are accused of managing their affairs not well at all.

I maintain that new legislation now is premature. The apparatus that Jay Keyworth established in 1984 and 1985 is healthy and functioning. The Biotechnology Working Group has delivered its report and recommendations to the Cabinet's Domestic Policy Council. The Council approved these on May 20, as a precursor for the President's review. None of this would have happened if OSTP's Biotechnology Science Coordinating Committee had not labored zealously under enormous pressure. Between the Biotechnology Working Group which is dedicated to policy considerations, and the OSTP's Biotechnology Science Coordinating Committee which is dedicated to scientific considerations, the Executive Office of the President has overseen the development of the new coordinated framework.

The BSCC was established to coordinate interagency review of scientific issues related to the assessment and approval of biotechnology research proposals, biotechnology product applications, and post-marketing surveillance. It serves as a coordinating forum to address scientific problems, share information, and develop consensus. It has promoted consistency in the development of federal agencies' review procedures and assessments. It will facilitate continuing cooperation on emerging scientific issues. Given these elements of its charter, the BSCC determined that its highest priorities were: to establish consistency among the agencies in the use of current scientific knowledge and to have agencies use comparable and rigorous scientific reviews.

Considerable time was spent in seeking adoption by the agencies of the same or very similar definitions of those genetically engineered organisms subject to regulatory review. Members eventually agreed on the definition of an intergeneric, or new, organism, and on the definition of a pathogen. As a result, they also agreed on what could be excluded from review under these definitions. The agencies adopted the scientific principles underlying the definitions in ways consistent with their legislation. They will use the definitions to identify levels of review for microbial products, or as factors to consider in reviewing product applications or research experiments.

The agencies will seek to operate their programs in an integrated and coordinated fashion; together they should cover the full range of plants, animals, and microorganisms derived by the new genetic engineering techniques. To the extent possible, responsibility for a product use will lie with a single agency. Where regulatory oversight or review for a particular product is to be performed by more than one agency, the policy establishes a lead agency, and consolidated or coordinated reviews. All of this will be clear to you in the forthcoming Federal Register Notice.

Unfinished business by the Biotechnology Science Coordinating Committee includes defining "release into the environment" and revisiting the concept of "containment," both physical and biological. When the committee completes its examination of these topics, once more I am confident that the agencies will consider this guidance in new rule-making or modifying regulations affecting genetically engineered organisms. Also, there remains the question of how to deal with those organisms that exchange DNA by known physiological processes. The public will have the chance to comment on the new definitions per se and their applicability to environmental release, to contained industrial large-scale operations, to food and food additives, drugs, medical devices, and other possible products.

In closing, I briefly return to "Leadership" in biotechnology. Leadership is needed from more than researchers and those who will develop products. All will lose or gain by the quality of leadership in the federal agencies, the congressional committees and their staffs, the trade associations, in short, all the stakeholders in biotechnology.

Congress can show leadership in the regulatory arena by giving the current statutes and authorities a chance to work within the administration's revised version of the Coordinated Framework for Regulation. Industry can show leadership by continuing its cooperative attitude with the agencies and responding substantively to the new Federal Register Notice. Researchers and government officials can show leadership by their insistence that modern molecular and cellular biology are as applicable to plant science and agriculture as they are to health science and pharmaceuticals production. The relative neglect of academic basic research in plant science in the totality of federal basic research policy and financing no longer can be justified. Al Wood emphatically would agree.

Closing Remarks

Kenneth R. Tefertiller
Vice President for Agricultural Affairs
Institute of Food and Agricultural Sciences
University of Florida
Gainesville, Florida

Several years ago, Dr. Al Wood had a dream. His dream has become reality this week at this International Symposium on Biotechnology. This symposium has served to bring us together to underscore the policy implications of this new science and place it in an international perspective.

Just in the last decade, biotechnology has begun to revolutionize the biological sciences. We are in the midst of a major revolution, a major pioneering effort. Like any pioneers, while we look to the future and discuss the implications of biotechnology, it is really not clear exactly where we will end up. Perhaps, looking back 20 years from now, we will realize just how much a time for change this was. However, we must remember that biotechnology will in the end probably enhance, not replace, the science that has always worked for us.

This symposium has provided us with a comprehensive overview of many of the critical issues with which we, as scientific and industrial leaders, must deal. For example, the public health and risk question is uppermost in many people's minds. However, we must also face the reality that new products must be field tested. They must be judged for their nutritional value. Accompanying this, of course, is a constant reminder that we must be committed to maintain the highest scientific standards of research, This will do much to allay public fears, and also enhance the public's perception of the new and often complex biotechnology concepts. We have successfully dealt with public welfare issues in the past and we will effectively deal with them now.

This conference has also identified some important missing links. Technical problems such as the shortage of scientists and skilled technicians must be faced on the university and the industrial level. The university will provide most of these human resources and there are, therefore, profound implications for the structure of college curricula and courses.

Biotechnology has served as a magnet, pulling together industry, government, and the universities. There is a growing awareness that traditional adversarial relationships between industry, government, and the universities must be mitigated. New linkages must and will emerge to facilitate and expedite technology transfer. We must also remember that business traditionally develops science and moves it into the marketplace.

Another important missing link is funding. We have heard a lot in this conference about the need for additional funding—especially in the plant sciences. We have heard comments from many different agencies and organizations. A common thread that runs throughout is that everyone seems to want to leverage everyone else. Everyone wants to provide the seed money but no one has offered the major capital that will be required to fund the commitment we have discussed. Someone, somewhere, must have enough faith in biotechnology to adequately fund it. Government, industry, and education must work together toward this end. Industry, however, cannot be expected to fund the basic research that must be done. This research must be performed by the universities who have always served as the "discoverer" and that role will continue and will grow in importance.

Everyone has a reason to make a commitment to biotechnology. Its promise is so powerful that it cuts across disciplines and across national boundaries. While we must be aware of the competitive advantage it offers, we should be careful not to overstate the competition between nations while understating the cooperative aspects and the welfare of all peoples.

We in Florida have an intense interest in biotechnology. Florida is the gateway to the tropics. We have tropical climate and tropical soils. We experience tropical human and animal health problems and the problems of human aging. We have a substantial load of plant pests. Our airports attract multitudes of tourists from all over the world. They also attract pests from other countries. We have thousands of miles of coastline and a fragile environment. Nowhere in the United States is there a greater concern for natural resource or environment problems than in Florida. Biotechnology offers us a promising tool to address these problems. The great hope in Florida is that biotechnology will enable all aspects of Florida's economy to remain competitive in an international setting while remaining environmentally compatible in a state that is growing increasingly urban.

APPENDIX
SYMPOSIUM PROGRAM

Biotechnology: Perspective, Policies and Issues

An International Symposium

June 1-4, 1986
University of Florida
Gainesville, Florida

Sponsors:
Institute of Food and Agricultural Sciences, University of Florida

IC^2 Institute, The University of Texas at Austin

RKG Foundation

Conference Committees

ADVISORY COMMITTEE

Dr. Donald R. Price, Co-Chairman
Dr. Kenneth R. Tefertiller, Co-Chairman
Mr. Scott Bailey
Dr. Orville G. Bentley
Dr. Norman Borlaug
Dr. Nyle Brady
Dr. George Kozmetsky
Dr. Robert Q. Marston
Mr. Robert P. Wynn
Dr. E.T. York

ORGANIZING COMMITTEE

Dr. John F. Gerber, Chairman
Mr. Dean J. Fenn
Dr. J.R. Kirkland
Dr. Jack McCown
Dr. Charles L. Niblett
Dr. Donald R. Price
Dr. Indra K. Vasil

PROCEEDINGS EDITOR

Dr. Indra K. Vasil

PROGRAM PLANNING COMMITTEE

Dr. Charles L. Niblett, Co-Chairman
Dr. Indra K. Vasil, Co-Chairman
Dr. Fuller Bazer
Dr. Frank Busta
Dr. James Cato
Dr. David Chynoweth
Dr. John R. Gander
Dr. L. Curtis Hannah
Dr. Peter E. Hildebrand
Dr. Tom Mabry
Dr. John T. Neilson
Dr. Harry Nick

CONFERENCE COORDINATORS

Lenie Breeze
IFAS, University of Florida

Ronya Kozmetsky, President
RGK Foundation

Biotechnology: Perspectives, Policies and Issues

An International Symposium

Major breakthroughs achieved in cellular and molecular biology during the last decade have made it possible to genetically modify plant and animal cells. These powerful new techniques offer opportunities of improved utilization of microbes, animals, and plants in human welfare. Current scientific and commercial developments in these areas have raised important new policy issues concerning the implications and commercialization of biotechnology.

Through this symposium we will examine the state of the art of biotechnology and focus on the issues of university/industry relations, economic opportunities, and ethical questions in the commercialization of biotechnology and the transference of these technologies throughout the world.

The symposium is designed to provide a forum for exchange among business, academia, and government in looking at the present and future scientific developments and beyond to the implications and the commercialization of biotechnology and related national and international policy issues. It will consist of four sessions:

SESSIONS I AND II will look at the state of the art of biotechnology in the areas of plant, animal, biomedical, marine, environmental, and bioengineering sciences. The sessions will examine the background, current scientific issues, and accomplishments, identifying specific products and the potential time frames for their availability and the constraints that exist in these areas.

SESSION III will examine new institutional developments from university, industry, and international points of view. Through the university perspective it will deal with relationships between universities and industry, including intellectual property rights, degree of involvement of faculty, joint and sole ventures, and the impact on graduate instruction. The industry perspective will look at industry/university relations, funding mechanisms, and the development, exchange and publication of scientific information and possible repercussions of manpower needs and demands. The international perspective will deal with the current status and future development and utilization of biotechnology in developed and developing countries with particular reference to property rights and technology transfer.

SESSION IV will deal with the biotechnology industry and commercialization perspectives. It will examine product development and delivery from the point of view of a well-established and successful biotechnology enterprise, it will look at legal issues involving biotechnology R&D and the establishment of corporations marketing biotechnology products. It will also address venture capital requirements for the biotechnology industry.

AGENDA

SUNDAY, June 1, 1986

2:00-6:00 p.m. REGISTRATION, University Centre Hotel

6:00-7:00 p.m. WELCOMING RECEPTION at the Home of President and Mrs. Marshall M. Criser, Jr.

MONDAY, June 2, 1986

7:30-8:30 a.m. CONTINENTAL BREAKFAST

8:00-10:00 a.m. REGISTRATION, University Centre Hotel

8:30 a.m. WELCOMING REMARKS

Mr. Marshall M. Criser, Jr., President
University of Florida

Mr. Ray Iannucci, Florida High Technology and Industry Council

Dr. Charles B. Reed, Chancellor, State University System of Florida

9:30 a.m. HEARING: United States Senate

The Honorable Lawton M. Chiles

Presentations invited from representatives of business, government, scientific institutions and universities to evaluate and recommend national policies with regard to the issues and questions related to the implications and the commercialization of biotechnology.

10:40 a.m. —Break—

11:00 a.m. HEARING RESUMES

12:00 noon LUNCH

Dr. George Kozmetsky, Director, IC[2] Institute, The University of Texas at Austin

1:30 p.m. HEARING RESUMES

2:40 p.m. —Break—

3:00 p.m. HEARING RESUMES

5:00 p.m. ADJOURN

6:30 p.m. RECEPTION

7:30 p.m. DINNER

TUESDAY, June 3, 1986

7:30-8:30 a.m.	CONTINENTAL BREAKFAST
8:00-10:00 a.m.	REGISTRATION, University Centre Hotel
8:30 a.m.	SYMPOSIUM OPENING REMARKS Dr. Donald R. Price, Vice President for Research University of Florida

SESSION I: ADVANCES IN BIOTECHNOLOGY: BREAKTHROUGHS AND BOTTLENECKS

8:40 a.m.	Chairman: Dr. Robert M. Johnson, Florida State University
8:45 a.m.	OVERVIEW OF BIOTECHNOLOGY Dr. Philip H. Abelson, Deputy Editor, *Science*
9:20 a.m.	IMPACTS OF BIOTECHNOLOGY ON AGRICULTURE: PLANTS Dr. Peter Day, Plant Breeding Institute, Cambridge
9:55 a.m.	—Break—
10:15 a.m.	IMPACTS OF BIOTECHNOLOGY ON AGRICULTURE: ANIMALS Dr. Roger M. Weppelman, Merck, Sharp and Dohme Research Laboratories
10:50 a.m.	PANEL DISCUSSION: PROBLEMS AND NEEDS IN BIOTECHNOLOGY RESEARCH Moderator: Dr. John Bedbrook, Advanced Genetic Sciences, Inc. Dr. Fuller Bazer, University of Florida Dr. Joachim Messing, Waksman Institute of Microbiology, Rutgers University Dr. Indra K. Vasil, University of Florida Dr. Philip H. Abelson Dr. Peter Day Dr. Roger M. Weppelman
12:00 noon	LUNCH Dr. Orville G. Bentley, Assistant Secretary for Science and Education, USDA

SESSION II: ADVANCES IN BIOTECHNOLOGY: ENVIRONMENT, HEALTH AND ENERGY

1:30 p.m. Chairman: Dr. Wayne Streilein, University of Miami

1:35 p.m. IMPACTS OF BIOTECHNOLOGY ON BIOMEDICAL SCIENCES
Dr. Thomas W. O'Brien, University of Florida

2:10 p.m. IMPACTS OF BIOTECHNOLOGY ON MARINE & ENVIRONMENTAL SCIENCES
Dr. Rita R. Colwell, The University of Maryland

2:45 p.m. —Break—

3:05 p.m. IMPACTS OF BIOENGINEERING ON BIOTECHNOLOGY
Dr. Alan S. Michaels, Alan Sherman Michaels, Sc.D., Inc., Boston

3:40 p.m. PANEL DISCUSSION: PROBLEMS AND NEEDS IN BIOTECHNOLOGY RESEARCH
Moderator: Dr. William J. Whelan, University of Miami

Dr. Jaime L. Frias, University of Florida

Dr. John P. Howe III, University of Texas Health Science Center, San Antonio

Dr. Larry L. McKay, University of Minnesota, St. Paul

Dr. Thomas W. O'Brien
Dr. Rita R. Colwell
Dr. Alan S. Michaels

5:00 p.m. ADJOURN

6:30 p.m. RECEPTION

7:30 p.m. DINNER
Dr. Robert Rabin, Assistant Director, Life Sciences, Office of Science and Technology Policy, The Executive Office of the President

WEDNESDAY, June 4, 1986

SESSION III: GLOBAL, UNIVERSITY AND INDUSTRY PERSPECTIVES

7:00-8:00 a.m. CONTINENTAL BREAKFAST

8:00 a.m. Chairman: Dr. Robert Q. Marston, President Emeritus and Professor of Medicine, University of Florida

8:05 a.m. EUROPEAN PERSPECTIVE
Dr. John Maddox, Editor, *Nature*

8:40 a.m. DEVELOPING WORLD PERSPECTIVE
Dr. Yongyuth Yuthavong, National Center for Genetic Engineering and Biotechnology, Bangkok, Thailand

9:15 a.m. —Break—

9:35 a.m. UNIVERSITY PERSPECTIVE
Dr. Robert Barker, Cornell University

10:10 a.m. INDUSTRY PERSPECTIVE
Dr. John Bedbrook, Advanced Genetic Sciences, Inc.

10:45 a.m. PANEL DISCUSSION: IMPLICATIONS FOR RESEARCH SUPPORT AND TECHNOLOGY TRANSFER
Moderator: Dr. Mary F. Clutter, National Science Foundation

Dr. Lawrence G. Abele, Florida State University

Dr. Anson R. Bertrand, U.S. Agency for International Development

Dr. John Maddox
Dr. Yongyuth Yuthavong
Dr. Robert Barker
Dr. John Bedbrook

12:00 noon LUNCH

SESSION IV: COMMERCIALIZATION ISSUES

1:00 p.m. Chairman: Mr. Robert P. Wynn, Deloitte Haskins & Sells, Ft. Lauderdale

1:05 p.m. PRODUCTS AND HEALTH CARE DELIVERY
Mr. Gary R. Hooper, Genentech, Inc., San Francisco

1:40 p.m. COMMERCIALIZING NEW PRODUCTS: PROBLEMS AND PERSPECTIVES
Dr. Cyrus M. McKell, NPI, Salt Lake City

2:15 p.m. CHANGING ROLE AND SCOPE OF VENTURE CAPITAL IN THE COMMERCIALIZATION PROCESS
Mr. Scott A. Bailey, Southeast Bank, N.A., Miami

2:50 p.m. —Break—

3:10 p.m. PANEL DISCUSSION: BUILDING THE INFRASTRUCTURE FOR BIOTECHNOLOGY IN FLORIDA
Moderator: Ms. Mary Helen Blakeslee, Baldwin, Rust and Dizney, Inc., Orlando

Dr. Ruth DeBusk, Receptor Molecules, Tallahassee

Mr. Larrick Glendening, Microlife Technics, Sarasota

Mr. Thomas S. Krzesinski, Florida Progress Corporation, St. Petersburg

Mr. Gary R. Hooper
Dr. Cyrus M. McKell
Mr. Scott A. Bailey

4:10 p.m. CLOSING REMARKS
Dr. Kenneth R. Tefertiller, Vice President for Agricultural Affairs, Institute of Food and Agricultural Sciences, University of Florida

4:30 p.m. SYMPOSIUM ADJOURNS